Jacques Ellul and the Technological Society in the 21st Century

Philosophy of Engineering and Technology

VOLUME 13

For further volumes:
http://www.springer.com/series/8657

Helena M. Jerónimo • José Luís Garcia
Carl Mitcham
Editors

Jacques Ellul and the Technological Society in the 21st Century

Springer

Editors
Helena M. Jerónimo
School of Economics and Management
Technical University of Lisbon
(ISEG-UTL) & SOCIUS
Lisbon, Portugal

José Luís Garcia
Institute of Social Sciences
University of Lisbon
Lisbon, Portugal

Carl Mitcham
Liberal Arts and International Studies
Colorado School of Mines
Golden, CO, USA

ISSN 1879-7202 ISSN 1879-7210 (electronic)
ISBN 978-94-007-6657-0 ISBN 978-94-007-6658-7 (eBook)
DOI 10.1007/978-94-007-6658-7
Springer Dordrecht Heidelberg New York London

Library of Congress Control Number: 2013939971

Printed on acid-free paper

Springer is part of Springer Science+Business Media (www.springer.com)

Contents

Chapter 1
Introduction: Ellul Returns

Helena Mateus Jerónimo, José Luís Garcia, and Carl Mitcham

Many nineteenth century thinkers, convinced of the Enlightenment premise that both nature and society were intelligible, and carried away by the growing prestige of the sciences, saw progress as a natural human development and believed that rational criteria guided societal choices. Biological evolution also appeared to provide a model for change applicable to history. An associated triumphalism in modernity dominated European popular culture until the outbreak of World War I and the post-war rise of dictatorial regimes. Yet even then a positive view of science remained largely intact. Even after World War II, the Shoah, saturation bombings of civilians, and the atomic destruction of Hiroshima and Nagasaki, the industrialization of science proceeded at an ever faster pace, assisted by an increasing involvement of state power. The United States science adviser Vannevar Bush (1945) went so far as to present post-World War II science as an "endless frontier" and font of social benefits in healthcare, economic development, and military defense.

In the midst of this enthusiasm for science and technology there was unease and insecurity in popular culture. In the middle of the twentieth century new genres of science fiction worry films such as *Invisible Monster* (1950), *Them!* (1954), and *Invasion of the Body Snatchers* (1956) together with the suspense message dramas of Alfred Hitchcock's *Rear Window* (1954) and *Vertigo* (1958) began speaking to

H.M. Jerónimo (✉)
School of Economics and Management of the Technical University of Lisbon (ISEG-UTL) & SOCIUS, Rua Miguel Lupi, 20, 1249-078 Lisbon, Portugal
e-mail: jeronimo@iseg.utl.pt

J.L. Garcia
Institute of Social Sciences of the University of Lisbon (ICS-UL), Av. Prof. Aníbal de Bettencourt, 9, 1600-189 Lisbon, Portugal

C. Mitcham
Liberal Arts and International Studies, Colorado School of Mines, Stratton Hall 301, Golden, CO 80401, USA

H.M. Jerónimo et al. (eds.), *Jacques Ellul and the Technological Society in the 21st Century*, Philosophy of Engineering and Technology 13,
DOI 10.1007/978-94-007-6658-7_1, © Springer Science+Business Media Dordrecht 2013

a growing concern in the public mind. Jacques Ellul was one who understood the unstable foundations and contradictions of this post-war moment, a period that was simultaneously optimistic and fearful. His intellectual journey was an attempt to understand the course of history in his own time, a process that took him beyond prevailing contemporary ideas and dogmas. Ellul was part of a twentieth century trajectory in thought that revisited the relation between philosophy and science, turning away from both epistemology and scientism to a questioning of scientific and technological culture. This questioning included a re-examination of the anthropological meaning of the technoscientific undertaking, of the responsibilities scientists and engineers acquire in attempting to master the worlds of nature and society, and of the metaphysical attitudes that ground any modern faith in science and technology. Along with such diverse thinkers as Edmund Husserl, Lewis Mumford, Hannah Arendt, Günther Anders, Hans Jonas, and Ivan Illich, Ellul was a pioneer in re-framing technology in moral problematic terms. Each argued in distinctive ways that modernity lacked the resources for understanding the power for good and evil unleashed by technoscience.

1

Jacques Ellul was born in the village of Pessac, near Bordeaux, France, on 6 January 1912 and died there on 19 May 1994, at the age of 82. His life therefore spanned virtually the whole twentieth century and its radical changes in society and ways of life. While a secondary school student, he met Bernard Charbonneau, with whom he was to have a lasting friendship and intellectual affinity, ranging from a shared interest in ecology to a common critique of the prevailing form of economic development and technological society. He studied law at the University of Bordeaux and began to read Karl Marx; having been brought up in the Calvinist and Augustinian traditions, he would later extend his interests to theology. During the 1930s, together with Charbonneau, he was part of the Personalist movement led by Emmanuel Mounier. He also made a brief effort at involvement on the Republican side in the Spanish Civil War. He married in 1937 and became Professor of Law at the Universities of Montpellier, Strasbourg, and Clermont-Ferrand. Under the Vichy regime he was expelled from the teaching profession and moved to a small village in the Gironde, where he worked with peasants, was an active member of the Resistance, and undertook formal theological studies. In 1943, he became Assistant Professor of Roman Law and History of Law and Institutions in the Faculty of Law at Bordeaux. From 1947 on he also taught at the Institute of Political Studies in Bordeaux. His lectures focused on the philosophy and economic thought of Marx and his successors and on the study of technics and propaganda. He remained in these posts until his 1980 retirement.

During his academic years Ellul constructed an increasingly broad body of work in the social sciences, theology, and public engagement, but the one we primarily focus on in the present volume is his seminal 1954 book *La Technique ou l'enjeu du siècle.* Since its publication many of the issues touched on there, from the threat of nuclear war and environmental deterioration to risks and globalization, have only increased in salience. Particularly from the time it was published in an "American edition" in 1964 as *The Technological Society*, with a foreword by sociologist Robert K. Merton, this book has encouraged a diversity of thinkers to address *Technique* or technology as a theme for critical reflection. Ellul's own study on this topic expanded in *Le Système technicien* (1977) and *Le Bluff technologique* (1988) – which with *La Technique* constitute a basic trilogy – as well as other books such as *Propagandes* (1962) and *Sans feu ni lieu* (1975). In his interdisciplinary reflections on history, politics, law, social life, and theology he repeatedly pursued such questions as: How does modern technique influence human beings? What is the hidden enigma in that which we call technique (or technics), and what is the reality of that which we call modern society? As his own words explain:

> *La Technique* [1954] studies society as a whole; *Propagandes* [1962] examines the technical means which change opinions and transform individuals; *The Political Illusion* [1969] is a study of how politics is transformed through being part of a technological society; and *The Metamorphosis of the Bourgeois* [1967] of how classes are transformed in a technological society. The two books on the *Revolution* [1969, 1972] question whether it is possible to have a revolution in a technical society. *Le Système* raises another issue: 'technique' as a system within a technical society; or, what does systems analysis teaches us about the phenomenon of technique? Finally, *L'Empire du non sens* [1980] is a study of how art is transformed by the technical milieu (Ellul 1981: 156).

Ellul used the French *technique* (German *Technik*, English *technics*) in a broad sense. He disagreed with a tendency to limit technique to particular technical devices, the most obvious of which are machines, and insisted on understanding it as a set of methods, rationally determined and aimed at effectiveness in some well-defined context. In this respect Ellul distinguishes between isolated technical operations and the technical phenomenon manifest throughout such operations in modern technics. In premodern or traditional technics any method remained embedded in its particulars whereas modern technics has become disembedded from and therefore able to be applied to particulars. Equating technics with technical knowledge in this way seems to be in line with the Ellulian understanding of *technique*, although it is not an identification Ellul himself makes. All human action requires knowledge, and technological knowledge is undoubtedly now one distinctive cognitive engagement with the world: knowledge that can be formulated in terms of an input–output analysis does not look beyond itself. It is a rational knowledge of means rather than ends (about which it is commonly argued there can be no rational knowledge, only opinions and preferences). Such input–output means knowledge, once the inputs and outputs are contextually specified, can be formulated precisely and this endows technological knowledge with the illusion of certainty. For Ellul, the intellectual character of the modern age is bound up with the sovereignty of technique, because human reason has come to identify itself with technological thinking. Remarkably, in the same year that Ellul published *La Technique* the Martin Heidegger's "Die Frage nach der Technik" (1954) appeared, arguing that "the essence of *Technik* is nothing *technikishe*" and for an understanding of modern *Technik* as a *Gestell* or framing of the world in terms of *Bestand* or resources. There are obvious affinities between the two analyses and both have been subject to similar criticisms for their abstract character. But there is a concreteness to Ellul's that frees it from the weaknesses of a thinking associated with National Socialism.

Recall briefly the seven concrete characteristics Ellul finds in the modern phenomenon of technology: rationality (*rationalité*), artificiality (*artificialité*), automatism of technical choice (*automatisme du choix technique*), self-augmentation (*auto-accroissement*), monism (*unicité* or *insecabilité*), technical universalism (*universalisme technique*), and autonomy (*autonomie*). Rationality references the fact that every adoption of technique entails some conscious analysis, usually of an input–output type. Artificiality describes the character of a world more and more the product of human construction such that humans themselves become responsible for an ever increasing proportion of the maintenance for the environment in which they live. Automatism in technical choice is present insofar as technical rationality takes on a more or less automatic character and is assumed to be "the one best way" to make decisions that themselves become calculations (e.g., in cost-benefit analysis). Self-augmenting growth emerges when technique reaches what economists once called the "take off" stage of economic growth, when growth becomes self-sustaining. Indivisibility denotes the way the components of technological systems become unified wholes acquiring a degree of independence as a technical milieu that paradoxically also requires constant attention and maintenance. Eternal vigilance is the price of artificial complexity. Technological universalism highlights both the tendency for technology to expand geographically, absorbing all countries, peoples and civilizations (through factors such as war, trade, transport, communications, and the export of technical labor), and its dominance over all fields and activities. In his description of technique, Ellul draws attention to the fact that it acts as much on the substance of the inorganic world (he cites the example of the atom; we could now mention nanotechnology) as on the organic (now in genetic and molecular, synthetic biology). The distinction between the born and the made is gradually subverted.

Characteristic autonomy, which partially incorporates some other concrete features, has been the most provocative and widely discussed of Ellul's key aspects of the technical phenomenon. Technology is autonomous in relation to economics, politics, morality, and religion insofar as these other social institutions find it increasingly difficult to exercise their independent forms of life. Just as in the European Middle Ages the church might have been described as autonomous insofar as it held sway over many other social institutions, so in the modern world technology appears to hold pride of place. Neither economic nor political priorities govern technological change: technology itself shapes other forms of social change. Although the particularities of technical change are influenced by entrepreneurs taking advantage of new affordances (as with such innovations as Google or Facebook, for instance), the deeper technical structures are less determined by external than by internal logics (Moore's law of increasing computing power, for example). As Ellul writes in one summary statement from a page early in *La Technique*: "Technique has become autonomous, creating its own devouring world, which is a law unto itself, denying all tradition" (Ellul 1954: 12). Although such language has been largely rejected in scholarly parlance in favor of arguments for social construction, for many high-tech workers there is something about it that continues to ring true. For instance, Kevin Kelly (2010), the founder of *Wired*, the original techno-glamour magazine, writes unabashedly about "what technology wants" and its autonomy.

Technological patterns and the direction of technological innovation over the last decades are broadly in line with the characteristics of technology as Ellul continued to observe them in *Le Système technician* and *Le Bluff tecnologique.* Consider the following selective examples: with regard to artificiality, technology increasingly dominates organic life through the increasing "technification" of biology and associated commercializations. A wide variety of synthesized organic substances are used today in a multiplicity of industrial applications, including in the sensitive areas of food and health. With regard to self-augmentation and monism, there is the field of "anthropotechnics," which is driving the construction of what one philosopher has called a "human park" (Sloterdijk 1999), or perhaps more aptly, a human zoo, in addition to the world of the genetic super- and bio-markets, of *babybusiness* and of liberal micro-eugenics. Technological convergence is part of the synergistic cross-fertilization of nanotechnology, biotechnology, information technology and new technologies based on cognitive science. In *Le Bluff tecnologique*, before turning to the domain of entertainment, Ellul put forward an idea that is the key to the forms of organization which structure our world: the "science-technology-commodity complex" (1988: 412). The same is being manifested in globalization (or *mondialization* in French) and the creation of a scientific-technological-trade complex. Originally in *La Technique* and then again in *Le Système technicien*, Ellul glimpsed the fact that modern technology has become synonymous with the world as a whole, because the influence of technological forces reaches the whole planet, so that the former historical situation in which civilizations followed different paths, changes to one in which all are on the same pathway, moving in the same direction, albeit at different points or stages.

Eighteenth and nineteenth-century prophets of technological civilization such as Henri Saint-Simon and H.G. Wells had imagined technology as a peaceful endeavour that would serve human purposes. Ellul's theories, worked out in the middle of the twentieth century, show us a technology associated at least as much with war, economic competition, planetary globalization of the market, and the power of the big corporations. For Ellul, technology, much more than capital, is the core element of modern civilization, and we have to recognize today that not only has technology acquired much greater power to shape and condition humanity, but that it has also merged with capital in an intensely dynamic fusion. The idea of the science-technology-commodity complex is a true picture of the system in which we live, in which science, research, and the university are all driven by the search for efficiency and placed at the service of the demand for even more technological innovation directed at the global market.

His illuminating and prophetic work on the emergence of the phenomenon of technology has acquired classic status among those who interpret the advanced societies of our age as inherently technological. The concept of "a classic" means that those who study and write about society today believe they can continue to learn from the work of Ellul. In many intellectual and academic circles *La Technique* was received as one of the most significant works to be read by anyone who wanted to understand what has been happening in the modern world. International recognition for Ellul began with the reception given to the publication of *The Technological Society* in the English-speaking world, followed by *Propaganda,* each work shedding light on the other. The Canadian philosopher George Grant, for instance, in his review of *The Technological Society* wrote, "Nowhere is Ellul clearer than in dealing with the great liberal chestnut that technique in itself is never wrong but only the use men make of it" Grant (1998 [1966]: 396). In the specific field of studies of technology and the technological society, Ellul's work lays down some fundamental criteria for debate. His work continues to be controversial while encouraging to networks and societies (such as the French Association Internationale Jacques Ellul and the U.S. based International Jacques Ellul Society) dedicated to discussing his legacy.

2

The year 2012 marked the centenary of Ellul's birth. The publication of a book in honor of this occasion is an opportunity to reflect once again on his thought and on the best ways of evaluating and honoring his legacy. In June 2011, a bilingual international conference was held at the *Instituto de Ciências Sociais* of the University of Lisbon (ICS-UL), Portugal, titled *Rethinking Jacques Ellul and the Technological Society in the 21st Century/Repenser Jacques Ellul et la Société Technicienne au 21éme Siécle*; the object was expressly to discuss Ellul's legacy. The essays now being published derive from that conference, by scholars of diverse nationalities – Canada, France, Portugal, Romania, South Korea, Spain, United Kingdom, and United States – who approached Ellul from diverse perspectives. Overall, they provide a lively exchange of interpretations on the technological society today, and testify to the continuing impact of Ellul's thought.

The book is divided into three parts. The first discusses Ellul's diagnosis of modern society, and addresses the reception of his work on the technological society, the notion of efficiency, the process of symbolization/de-symbolization, and ecology. The second analyzes communicational and cultural problems, as well as threats and trends in early twenty-first century societies. Many of the issues Ellul saw as crucial – such as energy, propaganda, applied life sciences and communication – continue to be so. In fact they have grown exponentially, on a global scale, producing new forms of risk. Essays in the final part examine the duality of reason and revelation. They pursue an understanding of Ellul in terms of the depth of experience and the traditions of human knowledge, which is to say, on the one hand, the experience of the human being as contained in the rationalist, sociological and philosophical traditions. On the other hand there are the transcendent roots of human existence, as well as "revealed knowledge," in the mystical and religious traditions. The meeting of these two traditions enables us to look at Ellul's work as a whole, but above all it opens up a space for examining religious life in the technological society.

The first essay evokes Ellul's most celebrated work of 1954. Carl Mitcham discusses why the book was so much more popular in the United States than in France or anywhere else. Going beyond the general critical background of thinkers about technology such as Spengler, Jaspers, Mumford, Ortega y Gasset, Giedion, Heidegger, and the radical American tradition of concern with nature as found in Emerson, Thoreau, Muir and Leopold, Mitcham believes that Ellul's popularity in the US was due to a chance affinity between his analysis and the experience of two distinct social groups: Christian social critics and political demythologizers, both of whom appropriated Ellul's ideas. The Christian social critics were involved with the Christian churches in the struggles of the civil rights movements and ecclesiastical contamination by racism. The political demythologizers were opposed to the myth of American exceptionalism, which prevailed even while admitting its errors in Vietnam.

Ellul's ideas cannot be taken as a closed system. Rather, his thoughts on modern society and rationalization should be compared with traditions such as the sociology of Max Weber. This is what George Ritzer does on the basis of his concept of the "McDonaldization of society." For Ritzer, the common factors in the "McDonaldization of society" (which seeks to enlarge on Weber's theory of rationalization) and Ellul's ideas on technique are the central role attributed to certain characteristics such as efficiency, predictability, calculation and control, and the weighing up of the irrational consequences they may have, such as dehumanization and disenchantment. However, a number of other factors separate him from Ellul, whom he considers to have a dystopian vision of the future. In Ritzer's view, Ellul's analysis could benefit from having a more refined and differentiated appreciation of technique, so as to incorporate the idea that some techniques are less of a problem than others or that there are some areas of life less subject to technique than others. This would avoid a reified vision of technique and would recognize man's key role in it – including that of contesting it.

The prevailing context of rationality in technological civilization, and its obsession with effectiveness, evidence, and univocity, disturbs and reduces the scope for symbols and symbolization. The technoscientific culture that dominates practically all domains of human existence reduces symbols to the level of signs, marginalizing symbolic language and affecting the whole of human culture. Starting from the idea that technical rationality produces irrational outcomes and that technical action, which is supposedly organized on the basis of objective concepts and means, has a significant symbolic dimension, Daniel Cérézuelle reflects on facets of cultural disorganization in the technological society of modern life to argue that the symbolic world which accompanies the process of technification and universalization of monetary relationships may weaken the anthropological foundation that hitherto made technification possible. We live under the "spirit of technicism," as he calls it, in a clear evocation of Weber. Modern life has a number of features that contribute to the erosion of our symbolic capital: the modern-day inflation in signs and images and the rapid changes taking place in the technical infrastructure; the monetization and commodification of modern economic life, which drains the life out of the non-monetary sphere, on which the reproduction of symbolic capital depends; the role of technoscience as a powerfully de-symbolizing social operator, which means that nothing remains intangible and everything is subject to change through the calculations of technical operations. Cérézuelle argues that there is an urgent need to demythologize this technicist or productivist spirit or imaginary.

The coexistence of the logics of symbolization and de-symbolization which are characteristic of technological development is also at the heart of Yuk Hui's essay. Using an Ellulian approach, in which the development of the technological system is a process of de-symbolization, and its principal dynamic the dialectical relationship between de-symbolization and re-symbolization by consumption, Hui sees an affinity with the ideas of Gilbert Simondon. Taking current information technology as his starting point, Hui suggests that we should go further in analyzing de-symbolization, because we are witnessing other forms of de-symbolization which go beyond mere re-symbolization by consumption: there is materialization through superabundant production and processing of data, which are now not just technical, but digital as well, giving rise to a digital milieu. While Ellul had identified the relevance of data processing as an extensively de-symbolizing force at the end of the 1970s, before the proliferation of the personal computer and the Internet, everything is now on a much larger scale. On the one hand, circuits have been created within a retentional system (which is also part of the technological system), and on the other humans have acquired the ability to mediate and anticipate. In other words, de-symbolization is also externalization, a process which the philosopher Bernard Stiegler has described as "tertiary retention." Through the analysis of these two aspects of de-symbolization, Hui seeks to update Ellul's concept of the technological system.

Wha-Chul Son proposes to analyze and interpret the notion of efficiency in Ellul's thought, and suggests we should activate what he calls "purpose driven technology," a new form of technology justified by its ends and not by efficiency. Despite the fact that the "efficiency principle" (EP) is one of the main elements of modern technology, Ellul did not pay much attention to it, particularly when compared to the concept of "autonomous technology." Son argues that the prevalence of the notion of efficiency in modern societies is based on the assumption that all elements can be controlled, including human elements, and that everything can be planned and measured. In this sense, the EP can be seen as the prototype of the "technological bluff," to the extent that it is used to justify any technological development whatsoever. The EP completes the autonomy of technique because, beyond effective efficiency in terms of input and output, it describes a situation in which people accept any device or activity provided that it is characterized as efficient. For Ellul, such assumptions were not only false, but also distorted the reality of the technological society and reduced the scope of personal freedom (by producing "non-freedom"). The "purpose driven technology" which Son puts forward tries to recover human initiative and control over technology, countering the increased autonomy of technology that derives from the EP.

Fashionable theories of "ecological modernization" are also based on the idea that efficiency-based management and confidence in technological development, market mechanisms and the State, can overcome the environmental crisis. Isabelle Lamaud reflects critically on this theory on the basis of Ellul's writings on ecology, a field in which he was highly influential and is regarded as having been a pioneer. Lamaud's analysis does not focus on the capitalist aspects of this theory; she suggests rather that in objectifying and technifying environmental issues, ecological modernization is an obstacle to the questioning of the modernist beliefs which sustain the myth of technical progress. Lamaud argues that the theory of ecological modernization is a kind of "technical ecology," a technical response to a problem which has itself been defined as technical, based on a belief that technique is neutral and the idea that technological development is the only way of dealing with the environmental crisis. The theory thus realizes one of Ellul's fears, that "environmental protection" would effectively not allow technological development to be questioned. In Lamaud's opinion, Ellul's ideas open up the possibility of a non-technical ecology, which is not necessarily anti-technology or technophobic, but that situates it within a framework of social and political concerns.

The second part of the book opens with an essay by Langdon Winner, which offers important insights on the main features of propaganda identified by Ellul, using the example of the popular American TV channel Fox News. Despite its publicity slogans, which advertise its objectivity and impartiality, Fox News frames all its alleged news in a right-wing perspective, which includes a mix of social conservatism, free-market, libertarian, traditionalist, fundamentalist and evangelical Christian, anti-black, anti-gay, anti-immigrant, American nationalist, militarist, and corporatist views. Fox is indifferent to its errors, distortions and lies, and occupies fourth place in the ratings. The consumers of propaganda, as Ellul described them, are not innocent receivers but active participants who seek out and even provoke the psychological action of propaganda. Democracy in modern societies depends on the use of propaganda to mobilize citizens to take part in political processes and as such paradoxically neutralizes those same citizens's original thoughts, civic deliberations and decision-making initiatives. Ellul pointed to the need for trust in direct experience and our own judgment on important social, economic, and political issues. Citizens should avoid pre-defined visions of reality offered up by media professionals, corporate managers, or the agents of any ideology. According to Winner, Ellul's advice here is necessary counsel for the future of democracy.

In a closely related analysis of contemporary society with a focus on cyberculture and the virtual world of global communications, Andoni Alonso considers three major topics in Ellul. One concerns the sacredness that has been acquired by the technocratic discourse of speed, while a second considers the possible means of resistance in the critical discourse generated within cyberculture by hackers or media specialists. Cyberspace and virtual reality are a magic realm for many scientists, some of whom even argue for a certain cyberspirituality, vindicating Ellul's observation that technology has become a new religion with its own imagery and theology. But this new religiosity ignores knowledge workers own psychosocial limitations, which in turn affects speed and acceleration. In a cyber-organized society, where the capitalism of knowledge is serviced by a new proletariat, computational technologies invade the whole of human life, and the question of speed, as Ellul foresaw, becomes a problem. With the replacement of organic time of attention, memory, and imagination by cybertime, work and leisure are progressively enmeshed in each other while both are undergoing their own fundamental transformations. According to Alonso, hackers and activists for free software represent the possibility of freedom in a world bound by the chains of institutions, corporations, and governments, and are turning into the "unseasonable thinkers" among whom Alonso classifies Ellul.

The resurgence of uncertainty, or unpredictability, as a result of the technological system is the focus of the essay by José Luís Garcia and Helena Jerónimo, who analyze the 2011 accident at the Fukushima nuclear power plant in Japan. After Chernobyl, this was the second most serious disaster in the history of nuclear power, one that took place in a country in the vanguard of technological progress. Behind the appearance of safety and control, the world is organized into technical macro-systems in which contingencies are camouflaged and subsumed into the category of calculable risks. Although nuclear accidents are usually classified as having extremely low probability, they are major and far-reaching events, and their consequences unknown, incalculable, and irreversible. Garcia and Jerónimo question the labelling of these events simply as "risks" and argue that this notion neglects everything which cannot be encapsulated in calculation formulae and underestimates the extent to which alleged gains in energy security are achieved in the shadow of possible catastrophe. On this basis they revisit Ellul's concept of foresight to stress the need for contemporary technological societies to live in a prudent manner, imagine worst-case scenarios, acknowledge that uncertainties are inescapable and realize that future catastrophes are the outcome of our own actions and are practically certain to occur.

Thinking about the real, potential consequences of technology and the issue of decision-making in a democratic context is the theme developed by Patrick Troude-Chastenet around the "Mediator" controversy. This medicine, produced by the French laboratory, Servier, was recommended for asymptomatic diabetes in people with problems of high cholesterol and triglycerides, and was also a powerful appetite suppressant. It was sold in France from the mid-1970s onwards. Studies gradually established that this medication caused heart problems, while at the same time the European Medicines Agency concluded that it was not effective in treating diabetes and that the risks involved outweighed the possible benefits. The medicine was withdrawn from sale in several countries many years ago, but in France it was only banned in 2009, with a death count by then running somewhere between 500 and 2,000. Troude-Chastenet compares this example of belated action by the French authorities to the "contaminated blood" case, the largest public health scandal in the 1980s and 1990s. Such cases offer clues on how to think about the decision-making process in pluralist democracies. For Ellul, authentic democracy has vanished and politics is better characterized by the rule of short-termism and necessity. In these particular cases, instead of increased protection for patients, there was a proliferation of control procedures and expert studies that diluted any personal responsibility. Troude-Chastenet reminds us that, for Ellul, proper political decision-making subordinates means to ends.

The rhetoric of economic necessity and of the inevitability of technoscientific management, used to justify the exploitation of the Alberta tar sands, the third-largest reserve in the world, is the theme of the analysis by Nathan Kowalsky and Randolph Haluza-Delay, who explain how this rhetoric overrides other values such as social stability, religion/spirituality, and sustainable development. Tar sands extraction is opposed by the indigenous peoples and by environmental organizations because of the environmental and social damage it causes, and defended by industry and both federal and provincial governments on account of its alleged economic benefits and the overriding need to ensure the well-being of the inhabitants. In a detailed description of the case, the authors show that both defenders and opponents of tar sands extraction base their arguments on the scientization of the topic. Even while approaching it from completely different angles, the discussion of environmental damage and public health issues surrounding the tar sands, the response to the request for a moratorium by civil society organizations, and the pastoral letter of a Roman Catholic bishop are all expressed in terms of technical rationality, thus corroborating Ellul's position that modern culture is embedded in a technological context.

Ellul explored the rationalist-philosophical and the religious traditions, stubbornly working to preserve the distinctiveness of each. The last part of the book focuses on this theme. Ellul's studies of religious experience in the technical society and the emergence of new forms of the sacred, myth, and religion have inspired many other thinkers. The essay by Fréderic Rognon examines the impact of Ellul's ethical and theological thought on French Protestantism. To this end, he seeks to shed light on Ellul's position in the theological and ecclesiastical context of contemporary French Protestantism and to outline the biographical and intellectual journey of some contemporary French theologians: Gabriel Vahanian, Jean-François Zorn, Olivier Abel, Antoine Nouis, Stéphane Lavignotte, among others. He concludes that Ellul's impact was due more to personal affinities than to a mass social phenomenon. But Ellul had a decisive influence on many individuals' intellectual and spiritual trajectories, extending far beyond the emblematic figures portrayed in this article.

Equally influential was Ellul's critique of the technological society to a group of theologians, engineers, and critics concerned about technology and social justice at the World Council of Churches in Amsterdam in the year 1948. Jennifer Alexander's essay shows how Ellul helped the group think of society in other than Marxist or capitalist terms. He rejected entirely the concept of planning inherent in both. The author analyses Ellul's speech and influence at that World Council of Churches meeting, in particular in the work of Committee III, and the papers drafted in preparation for the Amsterdam meeting. In the meetings and in the papers which circulated before the meeting, Ellul took up a radical position and was supported by a very large number of people in the ecumenical movement. Not all of Ellul's positions appear in the Committee's report, however, nor were they contained in the lecture he delivered to the Amsterdam Assembly. Despite the common concern with technique, there were differences among the Committee III members, and Ellul's vision differed from many others then circulating that criticized the technological society. Alexander argues that Ellul's contribution to the work of Committee III shows how his radical critique of the technological society has a theological foundation and contains insights into the theological features shared by cultures that have quite different productive and religious traditions.

Virginia Landgraf seeks to imaginatively establish a relationship between Ellul's thought and the Ten Commandments, focusing on the idea that the Decalogue defines the space in which life is possible. This reinterpretation of Ellul on the basis of his theological writings allows Landgraf to ask how people can fight back against the phenomenon of truth having collapsed into appearance. In other words, the collapse of human liberty, destiny and ultimate values into a reality expressed in terms of imaginary abstractions and a belief in power over objects which are seen as being manipulable, but which turn human beings into slaves of the reality they believe they control. Based on Ellul's distinction between truth and reality as "orders" having different characteristics, modes of transmission, logics and attitudes toward the world, Landgraf outlines two parallel readings of the Decalogue in the light of Ellul's theological and sociological writings. In the first, God is specifically named: He ensures that humans will live according to the dictates of the commandments. In the second, Ellul lets it be implicitly understood that people should resist the various ways in which truth collapses into reality. The author suggests that Ellul's interpretation of the Ten Commandments is of crucial importance for understanding the theory of the autonomy of technique. A significant part of the argument underlying this theory derives from the belief that mathematics provides definite, unequivocal results. Landgraf argues that the Ten Commandments shed light on a gap in Ellul's theory of autonomous technique, in that his argument that mathematical answers are indisputable derives not from the nature of mathematics itself, but from the belief that human beings, after the Fall, are envious of reality.

Gregory Wagenfuhr argues that Ellul's work is vital for understanding the modern world, for which the "post-modern" tag is inadequate. Drawing on Jean-Francois Lyotard, and linking his approach with Ellul's concept of technique, Wagenfuhr outlines a view of human life that revolves around the sacred, which integrates people into their milieu. Post-modernity becomes then a justificatory myth, an apparent religion, a diversity of legitimating narratives that disguise the true situation and serve merely to integrate individual persons into the technical milieu. For Wagenfuhr, the continuing use of the concept of post-modernity may turn out to be a "phenomenic error," because it diverts attention away from the truth of the current situation. It is like the phenomenic error that Ellul highlights in his book *The New Demons*, where he mentions that lack of awareness of the secularization of the modern world is one of the three "phenomenic errors" that have occurred in the entire history of Christianity.

Andrei Ivan compares Ellul and Peter Berger as non-conformist authors in relation to Christian revelation and society, based on their views on technology and the modern conscience respectively. Despite their different theological orientations, Ivan argues that a dialogue between the two thinkers provides useful guidelines for thinking about how the human mind has changed in the modern world. Both agree that Christian faith is being eroded. They start out from a common methodological premise, in that while Ellul criticizes the commonplaces of modern society, Berger questions that which the modern conscience takes for granted. For Berger, society is a social stage, a reality which has to be deconstructed, and this can be achieved by theology in its prophetic form. Tradition not only mediates religious experience; it also tames it. One way it does this is by adapting to the cultural background. This implies a "cognitive surrender," because the external challenge is internalized. Ellul disapproved of any Christian accommodation to the modern age, and was opposed to those who want to "Christianize" the state, society, its institutions, and morality. For him, Christians have made a gentlemen's agreement with culture, but that agreement was only possible because they allowed themselves to forget that Truth has been crucified by Reality.

3

We are pleased to have this opportunity to publicly acknowledge and thank the partners and institutions that have made this book possible, as well as the conference on whose proceedings it is based. In institutional terms, we are pleased to acknowledge support from the Institute of Social Sciences (*Instituto de Ciências Sociais*) of the University of Lisbon (ICS-UL), the School of Economics and Management (*Instituto Superior de Economia e Gestão*) of the Technical University of Lisbon (ISEG-UTL) and SOCIUS (Research Centre on Economic and Organizational Sociology) for the conference which took place in Lisbon in June 2011 on *Rethinking Jacques Ellul and the Technological Society in the 21st Century*. We wish as well to credit financial support from the Science and Technology Foundation (*Fundação para a Ciência e Tecnologia*) of the Portuguese Ministry of Science, the Luso-American Foundation (*Fundação Luso-Americana*), the Portuguese Society of Authors (*Sociedade Portuguesa de Autores*), the French Institute in Portugal (Institut Français du Portugal), the International Jacques Ellul Society, and the Association Internationale Jacques Ellul.

In personal terms, special thanks are due to Daniel Cérézuelle and Patrick Troude-Chastenet, two of the leading experts on Ellul, who served, along with the editors, as an advisory committee for the selection and assessment of the contributions to the present volume. Additionally, we are grateful to all authors who were patient and responsive to our continuous requests and to Justin Latici for assistance with editing. Finally, we wish to express appreciation to Peter Vermaas, Ibo van de Poel, and Ties Nijssen at Springer, the living agents behind the book series *Philosophy of Engineering and Technology*, for the welcome they gave this volume and for their outstanding management of the publication process.

References

Bush, Vannevar. 1945. *Science: The endless frontier*. Washington, DC: US Government Printing Office.
Ellul, Jacques. 1954. *La technique ou l'enjeu du siècle*. Paris: Armand Colin.
Ellul, Jacques. 1962. *Propagandes*. Paris: Armand Colin.

Ellul, Jacques. 1975. *Sans feu ni lieu: Signification biblique de la Grande Ville*. Paris: Gallimard.

Ellul, Jacques. 1977. *Le système technicien*. Paris: Calmann-Levy.

Ellul, Jacques. 1981. *À temps et à contretemps: Entretiens avec Madeleine Garrigou-Lagrange*. Paris: Le Centurion.

Ellul, Jacques. 1988. *Le bluff technologique*. Paris: Hachette.

Grant, George. 1998 [1966]. Review of Jacques Ellul's *The Technological Society*. In *The George Grant reader*, ed. Christian William and Grant Sheila, 394–398. Toronto/London/Buffalo: University of Toronto Press.

Heidegger, Martin. 1954. Die frage nach der technik. In *Vorträge und Aufsätze*, 13–44. Pfullingen: Neske.

Kelly, Kevin. 2010. *What technology wants*. New York: Penguin.

Sloterdijk, Peter. 1999. *Regeln für den Menschenpark. Ein Antwortschreiben zu Heideggers Brief über den Humanismus*. Frankfurt am Main: Suhrkamp Verlag.

Part I
Civilization of Technique

Chapter 2
How *The Technological Society* Became More Important in the United States than in France

Carl Mitcham

La Technique ou L'enjeu du siècle has an unusual history. The original French was published in 1954 and made scarcely a ripple in a cultural world dominated by Jean-Paul Sartre (*L'être et le néant*, 1943; *Saint Genet, comédien et martyr*, 1952; *Question de méthode*, 1957) and Albert Camus (*La peste*, 1947; *La chute*, 1956). Although *La Technique* received ten reviews, most were in periodicals associated with French Protestant intellectual life; only one appeared outside France, in Germany.[1] Somewhat surprisingly, the following decades witnessed translations into Spanish (1960), English (1964), Portuguese (1968), Italian (1969), and Japanese (1975). But most publishing houses were second tier and all non-English translations received little notice.

By contrast, the English-language "Revised American Edition," titled *The Technological Society*, appeared under the imprint of the prestigious publisher Alfred Knopf, graced with a foreword by the distinguished sociologist of science, Robert K. Merton. By 1967 at least six selections had been reprinted in other publications and the book had gone into paperback, where it has remained in print for almost 50 years. In no other version has the volume had such staying power. In mid-2011 on Amazon.com there were 19 reviews, of which 11 gave it the highest five-star rating. How is it that this rather abstruse book, loaded with French and European references, came to occupy such a prominent and persistent place in the American intellectual landscape?

[1] For a summary of reviews, see Hanks (2007: 317–320).

C. Mitcham (✉)
Liberal Arts and International Studies, Colorado School of Mines,
Stratton Hall 301, Golden, CO 80401, USA
e-mail: cmitcham@mines.edu

H.M. Jerónimo et al. (eds.), *Jacques Ellul and the Technological Society in the 21st Century*, Philosophy of Engineering and Technology 13,
DOI 10.1007/978-94-007-6658-7_2,

1 Background

The North American reception of Jacques Ellul's analysis of technology did not take place in a vacuum, but against a background of resistance to if not rejection of other more or less insightful critics who had preceded him. Thus it may be useful to begin a consideration of Ellul by referencing some of his predecessors.

Foremost among these predecessors is, of course, Karl Marx. As is well known, what might be termed Marx's phenomenology of economic commodities – as opposed to Hegel's phenomenology of ideas – was never given a sustained and serious reading in the United States. Marx's argument was rejected as placing too much emphasis on economics and as insufficiently appreciative of the socio-political dynamics of democracy and of technical ingenuity.

Marx does not, however, make technique a major independent theme of analysis. It was not Marx and the Marxists but the existentialists, including what might be called existentialist historians, who first broached technique as an issue for extended thematic consideration. Oswald Spengler, for instance, in the 1920s argued for commitment to the machine as the defining characteristic of modern Western civilization (Spengler 1922; see also Chase 1929). But more relevant to present purposes are five other studies:

- Karl Jaspers' *Die geistige Situation der Zeit* (1932);
- Lewis Mumford's *Technics and Civilization* (1934);
- José Ortega y Gasset's *Meditación de la técnica* (1939);
- Siegfried Giedion's *Mechanization Takes Command* (1948); and
- Martin Heidegger's "Die Frage nach der Technik" (1954).

The idea unifying these works is that in diverse ways modern technics has become technology which exists in tension with a truly human life.

German psychologist and philosopher Jaspers, for instance, having observed how human life has come to be seen as "the supplying of mass-needs through rationalized production on the basis of technical inventions," also notes that the new "life-order is perpetually troubled." In its effort to meet the real material needs of an increasing population, "[T]he mass-order brings into being a universal life-apparatus, which proves destructive to the world of a truly human life" – that is, one which undermines tradition and community (Jaspers 1932 [1955]: 29, 37).

Mumford, an American generalist, in *Technics and Civilization* – the first in a four-volume "Renewal of Life" series – identifies many positive transformative influences of technics, but argues that machine civilization needs to be transcended in a more truly life-centered technics. At the same time, there is no turning back. "*Until we have absorbed the lessons of objectivity, impersonality, neutrality, the lessons of the mechanical realm, we cannot go further in our development toward the more richly organic, the more profoundly human*" (Mumford 1963 [1934]: 363, italics in the original).

Like Mumford, Spanish philosopher Ortega sees *técnica* as an essential part of human nature – but only a part. Following his analysis of the existential foundations of *técnica* in human desire – desire not just for life but for a life of some particular sort – and of the human ability to stand outside any particular life as it might be historically given, in order to imagine other possibilities, Ortega notes that the history of technics is not a history of a univocal activity. The history of technique includes a trial-and-error technics of chance, a craft technics of the artisan, and a scientific technics of the engineer – the latter of which has created a special social problem. The modern engineer, by becoming absorbed in a perfecting of means can lose touch with the capacity to imagine ends. In Ortega's words, "To be an engineer and nothing but an engineer means to be potentially everything and actually nothing" (Ortega y Gasset 1945–1947 [1939]: 366). Modern technics runs the danger of undermining the imaginative life.

Swiss architectural and industrial historian Giedion returns more directly to the argument of Mumford, and elaborates in great detail the enormous post-Industrial Revolution expansion in the realms of mechanization. For Giedion this expansion has created an imbalance. Because "the human organism requires equipoise between its organic environment and its artificial surroundings" (Giedion 1948: 721), it is the task of the present to recreate the dynamic balance.

The German philosopher Heidegger sees *Technik* as more than a mere means, as a kind of revelation or truth; modern technics in fact constitutes the founding of a new way of being-in-the-world that in its elaboration tends to obscure a relation to Being. At the same time that modern *Technik* discloses beings as resource, technology itself is a manifestation of Being beyond resource. In Heidegger's words, "The essence of *Technik* is absolutely nothing technical" (Heidegger 1954: 13). Yet in the midst of the active and dominating presence of modern technics, it is increasingly difficult to accept or experience the ontological beyond that which technology brings into play in the world.

The appeal common to all five criticisms is to something larger or more encompassing than technology, against which technology should be measured. In contrast to the routine, the mechanical, the methodical that characterize technology, all five critics oppose something like life or a living eventfulness. Yet in all such criticisms, the primary manifestation of life (or Being) was tradition and community. Jaspers, Mumford, Ortega, Giedion, and Heidegger thus all constitute what can be called cultural criticisms of technology that are more general than but nonetheless related to the Marxist socio-economic critique of technology. Like the failures of Marxist and non-Marxist socio-economic criticism, when the new European cultural criticisms of technology began to become known in the United States between the 1930s and 1950s, they too received a largely negative response. Like Marxism, they were judged as ideological or unappreciative of what was really happening with technology in the New World.

Indeed, the only Marxism that achieved any significant purchase in the U.S. context was the culturalized version of Herbert Marcuse's *One-Dimensional Man* (1964), which appeared the same year as *The Technological Society*. As heir to the Frankfurt School Social Criticism of a new form of entertainment capitalism that colonized culture with products which reconciled the masses to political oppression, Marcuse sought through Freudian psychology sources for a liberation of the repressed. However, after the decline of the counter-cultural experience of the 1970s, his rationalization of the revolutionary potential of *avant garde* aesthetic expression, black power, and women's liberation received less than sustained attention.

In the United States what is primal is not an inherited tradition of culture and community or ancient cities with their established art and social orders that are threatened by technology; what is primal is nature – nature as wilderness – and the experience of new socio-cultural beginnings. As these new beginnings in society and culture became corrupted or failed to live up to their promise, the North American mind increasingly turned to wilderness as a fundamental good. In the United States it was the criticism of Ralph Waldo Emerson, Henry David Thoreau, John Muir, and Aldo Leopold and their appeal to life as manifested in uncontaminated natural wilderness that defined the most radical substrate of philosophical reflection.

Compare, for instance, the tradition of Emerson (1803–1882), Thoreau (1817–1862), Muir (1838–1914), and Leopold (1887–1948) with that of Marx (1818–1883) and V.I. Lenin (1870–1924). Both Marx and Thoreau judged the social order to be unjust and oppressive. But for Marx the response was to argue for socialist revolution, turning technology over to an oppressed class, the proletariat, in order to create a new society; for Thoreau it was to argue for a delimitation of technology and a protection of wilderness. Whereas Lenin reduced Marx's ideas to practice through the Communist Party, Muir did the same for Thoreau's idea by creating the Sierra Club.

The nature criticism of technology and what became the environmental movement has had a much more profound and lasting impact, especially in North America, than European socio-cultural criticisms of technology and associated socialist movements. But by the mid-1950s the challenge of radical environmentalism had become largely dormant. The initial success of the environmental movement in establishing a system of national parks, forests, and wilderness areas during the first half of the twentieth century had run its course. Partly as a result of such successes there remained much nature that was still untouched by human development, making it all the easier for World War II to distract the social imagination from environmental issues and the follow-on Cold War to capture the forefront of political attention.

Then in 1962 the publication of Rachel Carson's *Silent Spring* began to reactivate the distinctly North American tradition of environmentalism. Throughout the 1960s a revived environmentalism built up momentum not just to conserve some parts of nature from industrial or commercial development but also to protect nature from the secondary and side effects of development taking place outside parks, forests, and wilderness areas. This new environmental movement led, for instance, to legislative action with the National Environmental Protection Act of 1969 and the executive establishment of the Environmental Protection Agency in 1970. And perhaps in part because Ellul presented the challenge of technology as the replacement of a natural by a technical milieu, his analysis was given a reception in the United States it had not otherwise received. Indeed, because of an altered context, the North American reading tended to oppose technology to nature in a way that cannot fully be justified by Ellul's own perspective on our age.

But although an environmental sensitivity distinctive to the United States provides a partial background to the reception of Ellul's study of technique, this ideational opening remains insufficient to account for the full range of Ellul's North American influence. Among other significant contributions must be counted the emergence of an increasingly consumerist culture and the experience of the Vietnam War, both of which were given special character by technology and presented the United States in new and dramatic forms with yet other corruptions of its New World beginnings. Yet there is more to the unique U.S. reception of *The Technological Society* than can be explained by such contextual factors.

2 From *La Technique* (1954) to *The Technological Society* (1964)

During the same period in which Heidegger was formulating his ontological questioning of *Technik*, Ellul was developing a systematic analysis of *la Technique* as the most important societal phenomenon of the modern world. In English, both Heidegger's Technik and Ellul's Technique (with an anomalous capital T), become "technology."[2] According to Ellul, capital is no longer the dominant force it was in the nineteenth century; instead it is "technology," which he defines as "the *totality of methods rationally arrived at and* [aiming at] *absolute efficiency* (for a given stage of development) in *every* field of human activity."[3]

Indeed, it is Ellul's aim to offer for the twentieth century the same kind of orientation toward essentials once provided by Marx's *Das Kapital* (1867). As Ellul says in a later autobiographical reflection on the period during which he began studies that would culminate in *La Technique*: "I was certain... that if Marx were alive in 1940 he would no longer study economics or the capitalist structures but technology. I thus began to study technology using a method as similar as possible to the one Marx used a century earlier to study capitalism" (Ellul 1981: 155). Furthermore, all the work conceived during that period and eventually realized

---

[2]The issue of how the German *Technik* and the French *technique* (even with a lower case t) become the English "technology" is complex. See Schatzberg (2006). In the present context, which is focused simply on a book in English, it is acceptable to by-pass this complexity.

[3]This definition, as added in a "Note to the Reader" to the "American edition" of *La Technique*, reads in English: "The term *technique*, as I use it, does not mean machines, technology, or this or that procedure for attaining an end. In our technological society, *technique* is the *totality of methods rationally arrived at and having absolute efficiency* (for a given stage of development) in *every* field of human activity" (Ellul 1964: xxv, italics in the original). Unfortunately there exists no French original of this definition. Simply on the basis of meaning, however, it can be conjectured that the translation "having" in this context is slightly misleading; in light of related discussions it seems reasonable to substitute "obtaining as a result," "aspiring to," or "aiming at" (see, e.g., *The Technological Society*, pp. 11 ff. and pp. 79 ff., and *The Technological System*, chapter 1, "Technology as a Concept").

> was intended to be, with few exceptions, part of the detailed analysis of this technological society. For example, *La Technique* [1954] studies this society as a whole; *Propaganda* [1962] examines the technical means that serve to alter opinion and transform the individual; *L'Illusion politique* [1965] is the study of what politics becomes in a technological society; *Métamorphose du bourgeois* [1967] looks at the social classes in a technological society. My two books on revolution pose the question of what kind of revolution is possible in a technological society… And, finally, *L'Empire du non-Sens* [1980] is the study of what art becomes in the technological milieu (Ellul 1981: 155–156).[4]

La Technique provides the fundamental analysis by distinguishing between what Ellul calls "technical operations" and "the technical phenomenon." Chapter 1 gives a brief historical overview of how technical operations or technics are many, traditional, and limited by the diverse contexts in which they occur, whereas the technical phenomenon – or *la Technique* – constitutes that uniquely modern form of making and using artifacts that tends to incorporate into itself all other forms of human activity and thereby to dominate human life. Driven by a commitment encapsulated in the popular belief that "the solution to the problems of technology is not less but more technology," there arises the technical phenomenon or the comprehensive pursuit of efficiency; that is, "technique has taken over the totality of human activities, not only those of productive activity" (Ellul 1954: 2).

Chapter 2 lays out a "characterology" of this technical phenomenon or modern technology, and describes it as exhibiting (1) rationality [*la rationalité*], (2) artificiality [*l'artificialité*], (3) automatism [*l'automatisme*, self-directedness], (4) self-augmentation [*l'autoaccroissement*, self-supporting growth], (5) monism [*l'insecabilité*, indivisibility], (6) universalism [*l'universalisme*], and (7) autonomy [*l'autonomie*]. Having outlined this characterology, Ellul proceeds to show how it provides a basis for understanding the relations between technology and the contemporary economy (Chap. 3), state (Chap. 4), and what in English would be called the private realm (Chap. 5). Even medicine, education, sports, and entertainment, become subject to the input-output, cost-benefit analysis in search of "the one best way" to achieve results (Ellul 1954: 75).[5] The basic argument is that *technique* has become the organizing and dominating *logos* of contemporary society.

Ten years after its original publication, having gone out of print in France, *La Technique* appeared in English translation. Normally such a book would not have been translated. According to translator John Wilkinson, a professor of philosophy at the University of California, Santa Barbara, the unusual circumstances surrounding its English appearance are as follows (see Wilkinson 1970). In 1959, at the instigation of Aldous Huxley, Wilkinson with some colleagues began a reading-discussion group on Ellul's *La Technique*. About the same time, Robert Hutchins, recently retired president of the University of Chicago and founder of a new Center for the Study of Democratic Institutions in Santa Barbara, got pointed toward Ellul

---

[4]Ellul's tendency to overlook details is well illustrated by this passage which, in the original, incorrectly cites the titles of two of his own books. The bracketed dates have been added.

[5]The phrase, which comes from Frederick W. Taylor, is in English in the original. See Kanigel (1997).

when he asked Huxley about any book that might be suitable for Center attention.[6] As a result of this confluence, in 1962 Wilkinson became a staff member of the Center, which undertook to support the translation. This interest sparked Ellul to do a modest revision, which did not appear appeared in French until 1990.

The primary difference between the 1954 and 1990 French editions is a "Postface" and appendix. In the 1964 English translation, the "Postface" served as chapter 6, "A Look at the Future." (A 1990 appendix, "Réflexions sur l'ambivalence du progrès technique," was not included, and instead appeared separately as an appendix to Ellul's essay "The Technological Order" [Ellul 1963: 10–37]). The English translation further included two introductory notes by Ellul – one called "Note to the Reader" (dated June 1963) on the definition of technique, another "Author's Foreword to the Revised American Edition" – and an index. (It would have been useful if the French version had included the original French for both of these supplements, especially the "Note to the Reader.") The 1990 bibliography remained unchanged from that of 1954.

3 Reception of *The Technological Society* in the United States

The first review of *The Technological Society* appeared in *Library Journal* in September 1964. This was the 24th review of any Ellul book to appear in English. *The Presence of the Kingdom* (translated 1951) had been accorded 12 reviews, all mildly positive and by Protestants. *The Theological Foundation of Law* (translated 1960) had received 11 reviews. In these, Protestant reviewers supported Ellul's critique of natural law, while Catholic reviews argued that the critique failed. None of these reviews mentioned that by profession Ellul was a historian and sociologist, and in particular had initiated radical critical analyses of technology.

The first review of *The Technological Society* by Harold Fruchtbaum, a librarian at Harvard University, set a tone of the search for a "balanced" assessment. "This is a book many will not like, but it is a serious work that will have to be reckoned with" (Fruchtbaum 1964: 3175). The second review, by labor journalist Arnold Beichman in the *Christian Science Monitor*, agreed that *The Technological Society* was "a disturbing book," but primarily because of its intemperate language and "flawed reasoning" (Beichman 1964: 7). A third review, by political scientist Harry Howe Ransom in the *Saturday Review*, presented Ellul's book as a vivid warning (Ransom 1964: 48). These initial reviews thus staked out the basic positions to be maintained in one form or another in all 20 plus reviews that followed. Either *The Technological Society* was a difficult book that must be reckoned with, or it was totally wrong, not to say sinister in its mistakes. Or it was prophetic. The only serious attempt to consider Ellul's sociological analysis as sociology was one by Robert Theobald in *The Nation* (1964); a review that Ellul himself in a subsequent exchange found "conscientious and discerning" (Ellul and Theobald 1965: 567).

[6] Interestingly enough, the Ellul-Huxley connection had been suggested in the review of *La Technique* by Maurice Duverger, "Esope et les techniques," (*Le Monde,* 4 November 1954: 7).

The rather limited dozen reviews in the two or three years immediately following the publication of *The Technological Society* were complemented in the 10 years after publication by more than a dozen selected, anthologized reprints and about as many articles or sections in books that dealt with Ellul's sociological analysis. These articles actually began 2 years prior to publication, with a theme issue of the journal *Technology and Culture* on "The Technological Order." This special issue was the proceedings of a conference sponsored by the *Encyclopaedia Britannica* and held at the Center for the Study of Democratic Institutions; the same proceedings appeared as a book the following year (Stover 1963). The proceedings led off with a precis of his book by Ellul, followed by an extension of his argument. Without either endorsing or opposing Ellul's particular analysis, the majority of participants simply argued that technology deserved more scholarly consideration as a social phenomenon than it had heretofore been accorded.

Post publication of *The Technological Society*, in a more critical vein, Charles Silbermann in *The Myths of Automation* (1966) rejected what he termed Ellul's argument for "a technical take-over" as distorted and inconsistent (Silbermann 1966, especially chapter 6: 97–114). Similarly, political scientist Victor Ferkiss (1969), biologist Herbert J. Muller (1970), civil engineer Samuel Florman (1972), and computer scientist Thomas G. Donnelly (1973) all found Ellul to be fundamentally mistaken in his analysis. The Harvard University Program on Technology and Society (1964–1972), a major institutional effort funded with a $10 million grant from IBM, also took a distinctly negative attitude toward Ellul.[7]

Following up the early calls for serious consideration of Ellul's analysis, Harvey Wheeler, a colleague of Wilkinson's at the Center for the Study of Democratic Institutions, wrote a thoughtful piece on "Means, Ends, and Human Institutions" for *The Nation* (Wheeler 1967). In 1971 there also appeared the first articles on Ellul in social science and philosophy journals (Sklair 1971; Mitcham and Mackey 1971), and Ellul's analysis was referenced in Denis Goulet's *The Cruel Choice*, a critique of Third World developmental theory and practice (Goulet 1971). The basic positions adumbrated in the first three book reviews were thus extended over the next 15 years.

But given the unsympathetic background – and a preponderance of negative or weak reviews – the question arises as to why the book should have had such persistent influence, as indicated by its continued sales and a citation index that includes virtually every major discussion of technology.[8] A vague alliance with the resurgent

[7] See, for instance, the dismissive comment about Ellul as a writer of "pessimistic literature about technology" in Mesthene (1970: 22) (Mesthene was director of the Harvard Program.) The review of *The Technological Society* in Harvard University Program on Technology and Society, *Research Review* no. 6, "Technology and the Individual" (1970), is more neutral, although the analysis in the text of this research review argues that Ellul's position is not especially rational.

[8] Note, for instance, that standard introductions to philosophy and technology continue twenty years and counting later to find it necessary to mention Ellul if only to dismiss him. See, e.g., Ferré (1988: 12, 108–112, 130, and 141); Ihde (1993: 33–36, 44, 97–98, and 102); Pitt (2000: 87); and Dusek (2006: 27ff).

conservation movement was clearly insufficient. Ellul's own answer is that "both intellectuals and the public at large [in the U.S.] seized on my book because it described exactly what they were already... experiencing" (Ellul 1990b: xiv). But there were numerous other such books, as Ellul himself notes, so the question remains. What seems much more crucial is that in the United States, Ellul's work had two different kinds of institutional backing.

First was the Center for the Study of Democratic Institutions, which continued to promote the book throughout the 1960s. The *Center Diary* magazine (1963–1967) regularly featured articles on technology or Ellul.[9] Immediately after publication translator John Wilkinson produced a Center pamphlet titled *The Quantitative Society* (Wilkinson 1964). Following up the 1962 *Encyclopaedia Britannica* Conference, the Center did a second symposium in 1965 and published a proceedings pamphlet, *Technology and Human Values*, with contributions by John Wilkinson, Gerald Sykes, Dennis Gabor, Myron B. Bloy Jr., Martin Grotjahn, Theodore Roszak, and Bertrand de Jouvenel (Wilkinson et al. 1966).[10] Center staff and associates wrote about the book in non-Center publications,[11] and through the Center others were introduced to and encouraged to take it seriously, so that Ellul's book wound up being mentioned and considered by numerous other authors.

Second, and more adventitious, was the attraction to Ellul's book of a Protestant left. Here the work of Jim Holloway of the Committee of Southern Churchmen and editor of the journal *Katallagete: Be Reconciled*, based at Berea College in Kentucky, was particularly important. Unlike in Europe, where Ellul felt the Protestant establishment never took him seriously, in the United States there existed a counter-cultural Christian community that did take him to heart. The connection between radical, anarchist Christianity and radical sociology of technology was provocatively attractive to a number of people influenced by the civil rights and anti-nuclear movements. A special issue of *Katallagete* was

[9]The *Center Diary*, which began as a six page newsletter and ended up at 72 pages per issue was published once in 1963, four times in 1964, four in 1965, six in 1966, and three in 1967. It was succeeded by the *Center Magazine* (1967–1987). Virtually every single issue of the *Center Diary* included mention of Ellul or technology. Among the more prominent examples: Issue no. 10 (January 1966) featured two articles on a 1965 Center symposium on "The Technological Society." Issue no. 11 (April 1966) reprinted a note by Alfred A. Knopf from *The Borzoi Quarterly*, vol. 14, issue no. 4, saying that *The Technological Society* had in its first year sold almost 3,000 copies and that Knopf was now publishing a second Ellul translation, *Propaganda* (1965). Issue no. 17 featured a long article by John Wilkinson, "Futuribles: Innovation vs. Stability" (pp. 16–24) with extended references to Ellul. The succeeding *Center Magazine*, which developed a circulation of over 50,000, continued through the early 1970s to reference technology as a major issue.

[10]For a remark on the commitment to this conference, see Frank K. Kelly's institutional history (Kelly 1981: 275).

[11]See, for example, Scott Buchanan's review (Buchanan 1965: 821–823).

devoted to Ellul in early 1970 and published separately as a book later the same year. This was the first book to be devoted to Ellul's work, and included favorable articles by such writers as Christopher Lasch and Julius Lester.[12]

4 Jacques Ellul and Rustum Roy

This brief overview of responses can be complemented and deepened by a more detailed consideration of one particular manifestation of Ellul's significant North American influence: an exchange of letters that began the decade following publication between the author of *The Technological Society* and one of his American intellectual partisans. In the late 1970s the East Indian-American materials scientist, U.S. National Academy of Engineering member, and scientific establishment gadfly, Rustum Roy, wrote to Ellul to solicit his cooperation in "a new venture" as a co-editor of the *Bulletin of Science, Technology, and Society* (*BSTS*). Roy's initial letters in 1979 provided an enthusiastic sketch of the emerging interdisciplinary, activist, academic STS movement as well as a Christian house-church Protestantism which was, during that time, focusing on the challenges associated with technology, among other issues.

Ellul responded with interest and even enthusiasm:

> I have been very interested both in your letter and your projects [he wrote], and I admire the achievements of these STS programs! We are so much behind in France… I accept with great pleasure to participate in your effort....[13]

On another occasion, Ellul wrote that he was "very touched" by Roy's "good news" regarding "the consolidation and expansion of STS… and the opening of the Christian community" to the challenges posed by *la technique*.[14]

For those who knew Roy (who died in 2010, a decade and a half after Ellul) and now know as well the subsequent disheartening history of the STS activist movement – especially Roy's version of that movement – it is difficult not to be a bit surprised by Ellul's own apparent lack of skepticism. Indeed, as a result and with full encouragement from Roy, Ellul accepted the mantel of founding co-editor of the *BSTS* which, in a few years, became the official journal of the National Association for Science, Technology, and Society – now the International Association for STS (which currently exists in a state of suspended animation).

[12]The special issue of *Katallagete* was vol. 2 (3–4) published in Winter-Spring 1970. The book: Holloway (1970). Contents: Holloway, "Introduction"; Jacques Ellul, Letter; Holloway, "West of Eden"; Gabriel Vahanian, "Technology, Politics and the Christian Faith"; Christopher Lasch, "The Social Thought of Jacques Ellul"; Julius Lester, "The Revolution: Revisited"; Stephen Rose, "Wither Ethics, Jacques Ellul?"; William Stringfellow, "The American Importance of Jacques Ellul"; James W. Douglass, "On Transcending Technique"; James Branscome, "The Educational Illusion"; and John Wilkinson, "The Divine Persuasion: An Interview on Jacques Ellul."

[13]Jacques Ellul to Rustum Roy, 3 June 1979.

[14]Jacques Ellul to Rustum Roy, 15 September 1979.

The Ellul-Roy correspondence from 1979 to 1988, chronicles in part the declining fortunes of the STS movement. For instance, in a 1983 memo to Ellul and two other members of the *BSTS* editorial team (Steven Goldman and William F. Williams) soliciting their views on how best to continue the journal when the original publisher (Pergamon) was declining to do so, Roy wrote,

> We have come to a fork in the road in the life of BSTS. As you may already know, subscription numbers still remain low.... The 'paradigm' of specialization retains its dominance.[15]

To this Ellul wrote back, "You know that I admire STS very much and I am surprised there are so few subscribers."[16]

Over the course of their correspondence, Roy also repeatedly pressed Ellul to visit the United States: offering to cover all expenses, including, if necessary, passage by ship. Here Ellul did steadfastly decline, although primarily because of health issues and other responsibilities. Ellul never became part of the jet-setting scholarly community pioneered by Roy and others. Yet when Roy came up with two alternatives – a televised broadcast and/or video tapping for participation by Ellul in the first NASTS annual meeting – Ellul was quite receptive. He provided Roy an introduction to the various French television network options. And afterward repeatedly pressed Roy for assessments of how his virtual participation was received. If Ellul remained a non-participant in the jet-setting generation, it seems likely that he might well have joined a generation of skypers.

Among the interpretations that may be drawn from this Roy-Ellul story, one is that Ellul was seriously interested in the influence of his work in the United States and desirous of contributing to a movement for social change, such as that imagined in at least one strand of STS. It is modestly remarkable, however, that Ellul, who always described himself as a social scientist, never offered any sociological analysis of the problematics of his own influence: that is, especially, the contrast between his lack of intellectual traction in France as opposed to the United States; or that Ellul was never asked to be part of what eventually became the much more academic STS school in France – i.e., the form of STS known as Actor Network Theory and promoted by Bruno Latour and Michel Callon from the Ecole de Mines in Paris.[17]

Additionally, the Roy-Ellul exchange identifies two strands of U.S. influence: religious and political. Roy himself was the leader of a house church and an academic-social reformer who in both roles saw and sought to make alliances with Ellul's thinking. This distinction echoes the two institutional bases already referenced in considering how to account for Ellul's U.S. influence. The question, to restate, is this: What factor or factors contributed to making *The Technological Society* more popular in the United States – as illustrated by the Roy-Ellul

[15]Rustum Roy and Kathy Mourant to Jacques Ellul, Steve Goldman, and Bill Williams, 23 September 1983.

[16]Jacques Ellul to Rustum Roy, 30 September 1983.

[17]See, e.g., contributions by Latour, Callon, and others in Bijker and Law (1992).

exchange – than in France (or any other country)? The hypothesis is that Ellul's popularity in North America was a contingent result of the congruency of his analysis with the experience of two social groups and their quite distinctive American experiences.

Consider the following background and historical context: The United States, with a cultural conviction concerning "American exceptionalism" that can be traced back to the 1600s, had triumphed in World War II, primarily as a result of its technological prowess. It had out-built Germany, Japan, and all allied countries combined in the engineering and production of trucks, tanks, ships, military aircraft, and munitions. It had invented radar, the proximity fuse, long-range bombers, and the atomic bomb. After the war the industrial consumer economy roared into high gear – and in short order created the greatest wealth for the greatest number in human history. Utilitarianism reigned supreme. In such a context, American intellectuals largely rejected European social and cultural criticism of technology – especially any form of Marxism or Frankfurt School critical social theory – in favor of an almost unfettered optimism about techno-social progress. The only critics of technology who had even marginal purchase in the United States were activist heirs to the transcendentalist tradition of American culture.

At the same time, in the midst of abundance and optimism there began to emerge at the margins of public discourse a series of quite specific issues: fear of nuclear weapons, concern for the environment, uneasiness with corporate culture, concern for the civil rights of African Americans, and worries about consumer safety. Emblematic of the first, the Committee for a Sane Nuclear Policy (SANE) was founded in 1957; solidifying the second was the establishment of the Environmental Protection Agency (EPA) by Republican President Richard Nixon. The three other issues can be intellectually anchored in the following seminal books:

- William H. Whyte's *Organization Man* (1956),
- Martin Luther King Jr.'s *Stride Toward Freedom* (1958), and
- Ralph Nader's *Unsafe at Any Speed* (1965).[18]

But what is remarkable about such issues is how divided and disparate they remained. Few connections were made between them. There seemed no theory to explain them as instances of anything more fundamental than a countercultural movement.[19] Americans were left with their exceptionalism in tact and piecemeal social responses in the pragmatist tradition – except in two cases, where experience and frustration demanded deeper understanding. One of these involved the Christian churches and their critical social theoretical struggle in the civil rights movement. The other involved political activists against the Vietnam War who sought to demythologize an American democratic exceptionalist faith.

[18]For an extended presentation of this period, see Halberstam (1993).

[19]See, e.g., Marwick (1998).

5 Christian Critical Social Theorists

Consider, first, the case of the churches. The struggle with Jim Crow racism and violence against African Americans, first in the South and then across the United States, presented applied Christian theologians with the challenge of the social captivity of the churches. This was a phenomenon Ellul had analyzed theologically in the *Presence of the Kingdom* (French 1948, English translation 1951). It gave rise to what could be called "religious critical social theorists," especially among civil rights activists such as Will Campbell and Jim Holloway of the Committee of Southern Churchmen (CSC, founded 1964). Ellul's uncompromising theology as influenced especially by the thought of Søren Kierkegaard and Karl Barth provided a unique perspective on any and all tendencies to identify Christianity with worldly power. When in 1965 the CSC began publishing its radical evangelical magazine, *Katallagete: Be Reconciled*, one of its major efforts became the promotion of Ellul's work, including *The Technological Society*.

In the early 1970s, when I encountered Holloway, a professor in the Department of Philosophy and Religion at Berea College, a non-denominational school integrated from before the Civil War – and which had struggled for more than a century as an "out-post of free speech and abolitionist sentiment" in border-state Kentucky – he was firm in placing *The Technological Society* in theological perspective. For Holloway, demonic possession by racism and by technology were one and the same: possessions from which people could only be liberated by reconciliation with and in Jesus Christ. To emphasize his point, Holloway and Campbell were known to have celebrated a red-neck Easter in a manner particularly scandalous in the Southern Baptist world where they lived: by raising a glass of Kentucky Bourbon to proclaim "Jesus is Lord."

As Protestant theologian David Gill responded when queried about his own extensive evangelical engagement with Ellul, one manifested now in challenging techno-economic possessions in the corporate world of globalizing business:

> Ellul appealed to Barthians, to Anabaptists (like John Howard Yoder and my crew) because of his radical, nonviolent, biblical view of discipleship, and to many (not all) Evangelicals. I think Ellul felt he had more respect for his theology in the USA than in France, not just his sociology of technique.[20]

Yet Ellul's sociology of technology, precisely because of its Barthian roots, could be disseminated among Christian critical social theorists struggling as well with possession by racism.

6 Secular Political Demythologizers

Consider, as well, a second case, that of political demythologizing activists. The story of how *La Technique* was called to the attention of political demythologizers actually derives from a *Katallagete* interview with the translator, John

---

[20]David Gill, personal communication, 1 June 2011.

Wilkinson. Nevertheless, it was among the more demythologizing radicals on the margins of this scholarly community that Ellul's argument began to gain political purchase.

Among the most important political-theoretical readers in North American was Langdon Winner. According to Winner, when queried about his engagements, his own graduate school research began by making

> a trek to the Rand Corporation and to the Center for the Study of Democratic Institutions to see what 'futurists' and technology critics were pondering. I couldn't find anyone at the Center the day I dropped by [he writes], but I followed its projects and publications from afar. Years later I got to know W.H. 'Ping' Ferry who was one of the people who funded the Center.... Ferry was an old leftist, democratic socialist by inclination, heir to the Packard automobile fortune, a thinker in his own right....

Additionally, Winner observed how

> concerns brewing in the U.S. during the 1950s and 1960s about militarism, consumerism, environmental damage, media saturation of public life, the corporatization of higher education, etc. found a powerful focusing lens in Ellul's writing. His explorations of the dominance of 'la technique' seemed to connect a variety of more specific critiques (Vance Packard, Betty Friedan, Ralph Nader, Rod Sterling, New Left, etc.). Certainly, this came to include a good number of people in the student movement and counterculture.

He concluded that Ellul

> was especially good at denying anyone an intellectual or political safe haven. His consistent response to any path out of the wilderness was: That too is technique. This or that new gadget won't save you; neither will ethical professionalism, social movements, new forms of politics, etc. Different kinds of thought and action are needed.... The whole project of improving the world through amplified technological power was called into question. Among those who found this idea appealing were those who found the comforting ideas and policies of the liberal state inadequate in dealing with the major problems of the period.[21]

Not only did Ellul's analysis provide some unifying insight to multiple individual issues. As opposition to the Vietnam War intensified across the mid-1960s and into the mid-1970s – and seemed unable to utilize traditional democratic means to halt an unjust and increasingly technologized, devastating war – Ellul provided an effective demythologizing perspective on a searing American experience.

For many who came to political consciousness during that viscous period, the evil of which Americans in general continue to deny, Ellul offered a kind of insight absent in other quarters. When watching Errol Morris's documentary, *Fog of War* (2003) about the life of Secretary of Defense and un-prosecuted war criminal Robert S. McNamara, it is hard not to see McNamara's commitment to operations research analysis as illustrative of that banal dereliction of responsibility that characterizes the dark night of society Ellul sought to describe.

---

[21] Langdon Winner, personal communication, 22 May 2011.

7 Conclusion

The thesis, then, is that two fortuitous relationships contributed to making *The Technological Society* more popular and influential in the United States than in France or any other country. Christian critical sociologists and political demythologizers both enrolled Ellul's thought: the former to challenge the social captivity of the churches by racism and consumerism, the latter to expose the ideology of American exceptionalism that resisted admitting its evil actions in Vietnam.

Bill Vanderburg, a major Canadian interpreter of Ellul, when queried about this thesis, agreed that the two groups were "indeed the critically important channels through which Ellul's thought was sown in the cultural soil of the U.S." But he went on to suggest the need for demythologization precisely because the cultures of both France and the United States have "been profoundly transformed by new secular myths, which now 'possess' people individually and collectively." He noted further that Ellul's book with the English title *The New Demons* would have been more accurately translated as "The Newly Possessed."[22]

Finally, Daniel Cérézuelle, a French student of Ellul, suggests that because of its Catholic traditions and social hierarchies,

> one of the main obstacles to the reception of Ellul's thought [in France] was the pervasive idea that most social evils result from the search for profit (greed). Therefore, the response to these evils should be political (curbing the power of the wealthy) and institutional (in this case the State replacing the Church). That this explanation and this response are not sufficient can be better understood in a culture which puts more emphasis on personal responsibility (Protestant tradition).[23]

But what does this mean half a century later – as Ellul's argument has waned if not withered under the sophisticated academic attacks of counter theories regarding the social construction of technology?[24] Consider two speculative lessons:

First, in any contest between power and truth, truth does not succeed simply by virtue of its inherent persuasiveness. In the professional academic world of science, technology, and society scholarship – and in philosophy and technology studies more specifically – Ellul's influence is now largely dead as an illuminative moment, not just in France but in the United States as well. The impotence of insight in the contest with power – even academic power – is an old lesson, but one that has to be repeatedly re-learned. Recognizing how Ellul momentarily became prominent on the intellectual scene in the United States as a result of contingent social circumstances, should teach us to qualify all enthusiasms about the public influence of academic life. The current fate of political discourse in the United States about global climate change once again confirms this greater truth – which nevertheless does not absolve us of responsibility to continue to pursue such issues in the face of a self-interested and venal libertarianism.

[22] Bill Vanderburg, 31 May 2011, personal communication.

[23] Daniel Cérézuelle, 22 May 2011, personal communication.

[24] Note, for instance, the absence of any reference to Ellul in Olsen et al. (2009).

In footnote to this observation about the contest between truth and power, it is relevant to note that outside the powerful world of professional scholarship Ellul's insights still continue to find receptive audiences. In May 2011, *The Catholic Worker* newspaper published a long article on technology quoting extensively from Ellul (Walker 2011). In September 2012 the New Hope Catholic Worker Farm in Dubuque, Iowa, hosted its 3rd Annual Growing Roots Agronomic University devoted to "The Technological Question," also referencing ideas from Ellul.

But second, Ellul is not the final word. His one-time influence and popularity should not seduce us into rejecting Ellul's questions out of hand nor into thinking he has done it all – and all we need to do is become commentators on or elaborators of his work. There is no evidence that Ellul would have thought this to be the case, but both options nevertheless remain temptations in academia and out. In this respect we are fortunate that other equally insightful critics such as George Ritzer (1993) and Zygmunt Bauman (1998) have raised closely related challenges without any explicit reference to Ellul. More expansively, what Ellul's characterology of technology challenges us to do is to go beyond "all nuance and no news," to quote critic Adam Gopnik's complaint with regard to recent historical scholarship on the Inquisition.

> The pursuit of scholarly rigor too easily leads historians to erase any signs of the… imagination from their work. What is the historical imagination? It's simply the ability to see small and think big. Just thinking big leads you to Spenglerian melodrama and fantasy; just seeing small makes you miss history altogether while seeming to study it. After all, any significant change in human consciousness can be dissolved if you break it down into its individual parts, which are bound to seem contradictory or many-sided – you can dissolve anything by dissolving it (Gopnik 2012: 72).

What Gopnik says with regard to historical scholarship can apply even more strongly to significant representatives of sociological and philosophical work on technology. Ellul remains an alternative model, even for those who may disagree with his particular combination of seeing small and thinking big.

Acknowledgments For access to the correspondence between Rustum Roy and Jacques Ellul, I want to acknowledge Della Roy, Kathy Mourant, and Darryl Farber at Pennsylvania State University. Farber deserves further recognition and thanks for numerous conversations about the issues discussed in this article. Some issues were also discussed with Rustum Roy himself before his death in 2010. All translations from Ellul's French are my own.

References

Bauman, Zygmunt. 1998. *Globalization: The human consequences*. New York: Columbia University Press.

Beichman, Arnold. 1964. Is technique running us? *Christian Science Monitor* 24 September: 7.

Bijker, Wiebe E., and John Law (eds.). 1992. *Shaping technology/building society: Studies in sociotechnical change*. Cambridge, MA: MIT Press.

Buchanan, Scott. 1965. Untitled review. *George Washington Law Review* 33(3): 821–823.

Carson, Rachel. 1962. *Silent spring*. Boston: Houghton Mifflin.

Chase, Stuart. 1929. *Men and machines*. New York: Macmillan.

Donnelly, Thomas G. 1973. In defense of technology. *Christian Century* 90(3): 65–69.

Dusek, Val. 2006. *Philosophy of technology: An introduction*. Malden: Blackwell.

Ellul, Jacques. 1954. *La Technique ou l'enjeu du siècle*. Paris: Armand Colin.

Ellul, Jacques. 1960. *El siglo XX y la técnica* [Spanish translation of *La Technique*]. Madrid: Labor.

Ellul, Jacques. 1963. The technological order. In *The technological order*, ed. Carl Stover, 10–37. Detroit: Wayne State University Press.

Ellul, Jacques. 1964. *The technological society* [English translation of *La Technique*]. New York: Knopf.

Ellul, Jacques, and Robert Theobald. 1965. *The technological society*: A dialogue. *Nation* 200: 567–569

Ellul, Jacques. 1968. *A técnica e o desafio do século* [Portuguese translation of *La Technique*]. Rio de Janeiro: Paz e Terra.

Ellul, Jacques. 1969. *La tecnica, rischio del secolo* [Italian translation of *La Technique*]. Milan: Giuffré.

Ellul, Jacques. 1975. *Gijutsu shakai* [Japanese translation of *La Technique*]. Tokyo: Sugu Shobô.

Ellul, Jacques. 1981. *À temps et à contretemps*. Paris: Centurion.

Ellul, Jacques. 1990a. *La Technique ou l'enjeu du siècle*. Paris: Economica.

Ellul, Jacques. 1990b. *The technological bluff*. Trans. Geoffrey W. Bromiley. Grand Rapids: Eerdmans.

Ferkiss, Victor C. 1969. *Technological man: The myth and the reality*. New York: Braziller.

Ferré, Frederick. 1988. *Philosophy of technology*. Englewood Cliffs: Prentice-Hall.

Florman, Samuel C. 1972. Anti-technology: The new myth. *Civil Engineering* 42: 68–70.

Fruchtbaum, Harold. 1964. Untitled review. *Library Jourrnal* 89: 3175–3176.

Giedion, Siegfried. 1948. *Mechanization takes command: A contribution to anonymous history*. New York: Oxford University Press.

Gopnik, Adam. 2012. Inquiring minds: The Spanish Inquisition revisited. *New Yorker* 16 January: 70–75.

Goulet, Denis. 1971. *The cruel choice: A new concept in the theory of development*. Cambridge, MA: Center for the Study of Development and Social Change.

Halberstam, David. 1993. *The fifties*. New York: Villard.

Hanks, Joyce Main. 2007. *The reception of Jacques Ellul's critique of technology: An annotated bibliography of writings on his life and thought (books, articles, reviews, symposia)*. Lewiston: Edwin Mellon Press.

Heidegger, Martin. 1954. Die frage nach der technik. In *Vorträge und Aufsätze*, 13–44. Pfullingen: Neske.

Holloway, James Y. (ed.). 1970. *Introducing Jacques Ellul*. Grand Rapids: Eerdmans.

Ihde, Don. 1993. *Philosophy of technology: An introduction*. New York: Paragon House.

Jaspers, Karl. 1932 [1955]. *Die geistige Situation der Zeit*, 5th ed. Berlin: Walter de Gruyter.

Kanigel, Robert. 1997. *The one best way: Frederick Winslow Taylor and the enigma of efficiency*. New York: Viking.

Kelly, Frank K. 1981. *Court of reason: Robert Hutchins and the fund for the republic*. New York: Free Press.

King Jr., Martin Luther. 1958. *Stride toward freedom: The Montgomery story*. New York: Harper.

Marcuse, Herbert. 1964. *One-dimensional man: Studies in the ideology of advanced industrial society*. Boston: Beacon.

Marwick, Arthur. 1998. *The sixties: Cultural revolution in Britain, France, Italy, and the United States, c.1958-c.1974*. Oxford: Oxford University Press.

Mesthene, Emmanuel G. 1970. *Technological change: Its impact on man and society*. Cambridge, MA: Harvard University Press.

Mitcham, Carl, and Robert Mackey. 1971. Jacques Ellul and the technological society. *Philosophy Today* 15(2): 102–121.

Muller, Herbert J. 1970. *The children of Frankenstein: A primer on modern technology and human values*. Bloomington: Indiana University Press.

Mumford, Lewis. 1934 [1963]. *Technics and civilization*. New York: Harcourt Brace and World.

Nader, Ralph. 1965. *Unsafe at any speed: The designed-in dangers of the American automobile*. New York: Grossman.

Olsen, Jan Kyrre, Evan Selinger Berg, and Søren Riis (eds.). 2009. *New waves in philosophy of technology*. New York: Palgrave Macmillan.

Ortega y Gasset, José. 1945–1947 [1939]. Medicación de la técnica. In *Obras Completas*, vol. 5, 317–375. Madrid: Revista de Occidente.

Pitt, Joseph C. 2000. *Thinking about technology: Foundations of the philosophy of technology*. New York: Seven Bridges Press.

Ransom, Harry Howe. 1964. Man and method: Which is master? *Saturday Review* 47: 48.

Ritzer, George. 1993. *The McDonaldization of society: An investigation into the changing character of contemporary social life*. Newbury Park: Pine Forge Press.

Schatzberg, Eric. 2006. *Technik* comes to America: Changing meanings of technology before 1930. *Technology and Culture* 47(3): 486–512.

Silbermann, Charles. 1966. *The myths of automation*. New York: Harper and Row.

Sklair, Leslie. 1971. The sociology of the opposition to science and technology, with special reference to the work of Jacques Ellul. *Comparative Studies in Society and History* 13(2): 217–235.

Spengler, Oswald. 1922. *Der Untergang des Abendlandes: Welthistorische Perspektive*. Munich: C.H. Beckshe.

Stover, Carl F. (ed.). 1963. *The technological order: Proceedings of the encyclopedia Britannica conference*. Detroit: Wayne State University Press.

Theobald, Robert. 1964. The house that *Homo sapiens* built. *Nation* 199: 249–252.

Walker, Ted. 2011. Reflections on technology. *Catholic Worker* 78(3): 4–5.

Wheeler, Harvey. 1967. Means, ends, and human institutions. *Nation* 204: 9–16.

Whyte, William H. 1956. *The organization man*. New York: Simon and Schuster.

Wilkinson, John. 1964. *The quantitative society or, what are you to do with noodle?* Santa Barbara: Center for the Study of Democratic Institutions.

Wilkinson, John. 1970. The divine persuasion: An interview on Jacques Ellul with John Wilkinson. *Katallagete* 2(3–4): 57–65.

Wilkinson, John, et al. 1966. *Technology and human values*. Santa Barbara: Center for the Study of Democratic Institutions.

Chapter 3
The Technological Society: Social Theory, McDonaldization and the Prosumer

George Ritzer

This volume gives me the welcome and highly useful opportunity to address Jacques Ellul's (1964 [1954]) classic work, *The Technological Society*,[1] from the context of my work on social theory, the McDonaldization of society, and the age of the prosumer. I have found much of utility in the book and also much to criticize. On the positive side, Ellul's work on technique is very useful in thinking about the principles of McDonaldization and in speculating on the coming age of the prosumer. While the key idea of technique continues to be quite useful, Ellul's theoretical approach has its weaknesses and it has not worn well as a general theory.

1 As Social Theory

Written at the height of the modern era, it is not surprising that *The Technological Society* offers a very modern theory. In fact, it demonstrates in extreme form all of the limits of such theory (see below for discussion of several of them). The limits of modern theorizing (Bauman 1991, 1992) from a postmodern perspective were to be pointed out a few decades later, ironically almost always by other French theorists (e.g. Lyotard 1984 [1979]).

Even within the context of modern theory, there are problems with Ellul's approach, most notably his failure to deal with the work of highly relevant predecessors, especially Max Weber, whose ideas had much in common with his own (among the others who have noted this are Hayim 1978; Menninger 1981; Maley 2004). This failure will become clear when we discuss the relationship,

[1]Note: I focus only this key work and not the entirety of Ellul's output.

G. Ritzer (✉)
Department of Sociology, University of Maryland,
2112 Art-Sociology Building, College Park, MD 20742, USA
e-mail: gritzer@umd.edu

H.M. Jerónimo et al. (eds.), *Jacques Ellul and the Technological Society in the 21st Century*, Philosophy of Engineering and Technology 13,
DOI 10.1007/978-94-007-6658-7_3, © Springer Science+Business Media Dordrecht 2013

especially the similarities, between Ellul's ideas and Weber's, as well as my thinking on McDonaldization (Ritzer 2011) since the latter relies heavily on Weber's theory of rationalization.

Ellul and Weber are both interested in the means-ends relationship. Weber's work on both action and rationality is based in his concern for the means-ends relationship. In terms of action, means-ends rational action, in contrast to Weber's other types of action (value, affectual, traditional), is based on the search for the most direct means to any given ends. In terms of rationality, the key contrast for Weber is between practical (involving the pragmatic and self-interested choice of means to ends), substantive (values guiding the choice of means to ends), and formal rationality (where the choice of means to ends is guided, if not determined, by universally applied rules, laws and regulations). Ellul is similarly interested in the means-ends relationship as is clear in his definition of technical operation as including "every operation carried out in accordance with a certain method in order to attain a particular end" (1964 [1954]: 19). He was critical of the fact that in the pre-modern era there "was no great variety of means for attaining a desired result, and there was almost no attempt to perfect the means which did exist" (1964 [1954]: 67). The development of a variety of means to an end and the perfection of those means, even the finding of the best possible means, lies at the heart of modern technological society.

In a number of places Ellul links technique to Weber's favorite topic – rationality. For example, he argues that, "in the nineteenth century, society began to elaborate an exclusively *rational* technique" (1964 [1954]: 73, italics added). Later, he lists rationality – as well as "artificiality" – as the "essential characteristics of today's technical phenomenon" (1964 [1954]: 78). In fact, he accords rationality pride of place in his discussion of these characteristics. However, he chooses not to elaborate much on rationality – and artificiality – because "of their obviousness" and because they are emphasized by the "best authors" (1964 [1954]: 79). One wonders why the latter is in quotation marks and why those authors (especially Weber) are not even mentioned, let alone discussed. His unwillingness to deal with rationality is rendered even more incomprehensible by the fact that a variety of other theorists, most notably Karl Mannheim (1936 [1929], 1940 [1935]), have seen fit to devote considerable energy to developing the concept further (see also Kalberg 1980).

Not only is Ellul concerned with rationality, at least as it relates to technique, but he also has a dystopian view of the future that uses phrases very similar to those of Weber, especially the idea of the "iron cage" of rationality (Mitzman 1971 [1969]). For example: "Here ends the long *encirclement* of men by technique" (Ellul 1964 [1954]: 387, italics added); "*a ring of iron* with which technique surrounds [people]" (1964 [1954]: 415, italics added); and "[e]nclosed within his artificial creation, man finds that there is 'no exit'" (1964 [1954]: 428). With a similar concern for rationality and a similarly dystopian view of the future (although Ellul's is much more pronounced than Weber's), one wonders why Ellul apparently did little more than offer passing references to Weber. Their intellectual relationship to one another is even more striking when, in the context of a discussion of McDonaldization, we look at the basic dimensions of rationalization (McDonaldization) and see that they are prominent not only in Weber's thinking, but also Ellul's.

2 McDonaldization

The McDonaldization thesis involves an effort to apply, expand and further develop Weber's theory of rationalization. It brings Weber's theory up-to-date, at least until the late twentieth century, by arguing that it is the fast food restaurant and not the bureaucracy, as Weber contended, that is the paradigm of the process of rationalization. It is worth noting that with the further passage of time and changes in technology, other contenders for the paradigmatic role within the process of rationalization emerge. For example, given the rapid and extensive emergence of the computer and other digital technologies in the last few decades, it is possible to argue that eBay is now the paradigm of what might be termed "eBayization" (Izberk-Bilgin and Ahuvia 2011). While this is a defensible argument, it is the case that eBay is, at base, McDonaldized; it operates in accord with same basic principles (see below) as McDonald's. Of course, it could also be argued that *both* McDonald's and eBay are rationalized and that McDonaldization and eBayization are part of Weber's process of rationalization. At the minimum, what both of these phenomena and ideas point to is the fact that rationalization has continued to exist and extend its reach; it has spread into domains that Weber could not have even envisioned.

The McDonaldization thesis shifts the Weberian focus from other types of organizations, especially those that focus on production (e.g. the factory), to those that are concerned with consumption. They are the real heart of contemporary developed societies, especially in the US. McDonaldization is defined by its principles (see below) and it is argued that those principles are coming to affect more and more sectors of society (e.g. the church [Drane 2008 [2001]], the university [Hayes and Wynyard 2002]) and more and more parts of the world. In terms of the latter, it has come to be seen by many (see O'Byrne and Hensby 2011) as a theory of globalization even though that was, at best, only implicit in my original work on this topic.

A number of critiques of the "McDonaldization thesis" (Ritzer 1997) have revolved around the fact, observed by many (especially Watson 2006 [1997]), that there is variation around the world in the food served in McDonald's (and other McDonaldized chain restaurants). The argument is made that if the process of McDonaldization really existed, the food, like everything else, would be the same throughout the world (although I never argued it would). In the face of such criticisms, I often counter, "It's not about the food, stupid." Rather, what defines McDonald's, as well as other McDonaldized organizations, is the set of principles, the system, by which they operate. It is that system, those principles, that are (largely) homogeneous throughout the world and that define McDonaldization. This is clear in Ram's (2004) study of McDonald's and other fast food restaurants in Israel and in his argument that what defines them is "structural uniformity." Similarly, Bryman (2004), who develops a similar model of Disneyization, recognizes that McDonaldization (and Disneyization) are about systems and the principles that undergird them. In my view, the fundamental operating procedures remain essentially the same everywhere in the globe. Even a critic such as Watson (2006 [1997]) quotes the managing director of McDonald's in Singapore as saying that what McDonald's sells is *not* a product, but rather a system.

What defines McDonaldized systems in Ellul's terms is technique; the techniques by which they operate are codified and are applied, more or less uniformly, throughout the world. This view of the essence of McDonaldization fits well with Ellul's definition of technique:

> Technique is a means with a set of rules for the game. It is a 'method of being used' which is unique and not open to arbitrary choice; we gain no advantage from the machine or from organization if it is not used as it ought to be. There is but one method for its use, one possibility. Lacking this, it is not technique (1964 [1954]: 97).

In these terms, McDonaldization involves "rules for the game" of operating such systems, methods for operating and using them, methods that are unique to them, and the limiting or complete elimination of arbitrary choice within them. These systems succeed when they are operated in accord with the rules. In fact, there is only one way for them to operate in accord with the basic principles.

While this is the point of greatest overlap between Ellul's thinking on technique and mine on McDonaldization (as well as Weber's on rationalization), there is also a very profound difference which gets at what I consider the fundamental flaw in Ellul's thinking. We will get to this below, but before we do, we need to outline the basic principles of McDonaldization and discuss how closely they resemble basic principles outlined in Ellul's work on technique.

The first dimension of McDonaldization is efficiency and it is clear that Ellul associates this with the technique he sees triumphant in the contemporary world. Early on, he argues that technique "is efficient and brings efficiency to everything" (1964 [1954]: 5). As technical forms emerge over time, they "are not necessarily more complicated than the spontaneous ones, but they are *more efficient* and better adapted" (1964 [1954]: 20, italics added). The following lengthy quotation yields much insight into Ellul's thinking on the relationship between rationality, technique and efficiency:

> The intervention of rational judgment in the technical operation has important consequences. Man becomes aware that it is possible to find new and different means. Reason upsets pragmatic traditions and creates new operational methods and new tools; it examines rationally the possibilities of more extensive and less rigid experimentation. Reason in these ways multiplies technical operations to a high degree of diversity. But it also operates in the opposite direction; it considers results and takes account of *the fixed end of technique – efficiency*. It notes what every means devised is capable of accomplishing and selects from the various means at its disposal with a view to securing the ones that are the most efficient, the best adapted to the desired end. *Thus the multiplicity of means is reduced to one: the most efficient.* And here reason appears clearly in the guise of technique (1964 [1954]: 20–21, italics added).

In addition, he argues that the individual is "subordinate to the search for efficiency" (1964 [1954]: 74), or later, "technique has only one principle: efficient ordering" (1964 [1954]: 110), and still later "the end of technique is efficiency and rationality" (1964 [1954]: 201).

The second dimension of McDonaldization is predictability, and Ellul has much less to say explicitly about that than efficiency. However, he does say: "Technique requires predictability and, no less, exactness of prediction" (1964 [1954]: 138).

Then there is calculability, or the emphasis on quantity rather than quality. The strongest discussion of this dimension is, not surprisingly, in the section devoted to the economy where Ellul discusses, among other things, statistics, the application of mathematics, and the employment of accounting procedures. These procedures and those who use them are part of a larger change in society in the direction of an emphasis on greater calculability whereby "[e]xact quantities, weights, and times must be fixed" (1964 [1954]: 166). Of statisticians, Ellul says that there has been "change in the state of mind of the statisticians themselves. They are immersed in a 'statistical atmosphere' and comply with the *quantitative and numerical practices of the modern world*" (1964 [1954]: 163, italics added). In the context of a discussion of public opinion analysis, Ellul discusses the qualitative elements – "things hitherto unmeasurable" (1964 [1954]: 168) – that came to be transformed into that which is measurable, or lost altogether because they cannot be measured. As Ellul puts it: "Whatever cannot be expressed numerically is to be eliminated from the ensemble, either because it eludes enumeration or because it is quantitatively negligible" (1964 [1954]: 168). This not only reflects an emphasis on quantity, but also a corresponding lack of interest in quality because quality is, well, qualitative and therefore not able to be measured. More generally, from the point of view of both technique and McDonaldization, it is not just the economy that has come to be dominated by this calculating state of mind; all aspects of society have come under its thrall.

The fourth dimension of McDonaldization is control, especially through the substitution of non-human for human technology. This is implicit in Ellul's comparison to the pre-technological society in which "the accent was on the human being who used the tool and not on the tool he used" (1964 [1954]: 68). Or later, "[A] new machine of great productive power put into circulation 'releases' a great quantity of work; it replaces many workers" (1964 [1954]: 103). Man, to Ellul, is subject to "error and unpredictability" and therefore must "inexorably" be controlled and even removed from any decisive operations in contemporary technologies (1964 [1954]: 136). One of his strongest arguments on control is the idea that in technological society man "resembles a slug inserted into a slot machine; he starts the operation without participating in it" (1964 [1954]: 135). (Interestingly, this is another similarity with Weber who discusses rationalized law as resembling a slot machine [Ritzer 1975].) Later, "[b]ecause of the autonomy of technique, modern man cannot choose his means any more than his ends" (1964 [1954]: 140). While this is another point of similarity between Ellul and the theory of rationalization and McDonaldization, his position here also demonstrates another important weakness in his work, his tendency to reify technique it "pursues its own course" (1964 [1954]: 135) as well as the process by which technique comes to operate automatically without human inputs. Again, we will defer this discussion until the next section.

The most important aspect of my argument on McDonaldization, and Weber's on rationalization, is the irrationality of rationality. That is, rational, McDonaldized systems inevitably spawn a series of irrational consequences such as dehumanization and disenchantment. This is the critical center of the theory of rationalization/McDonaldization and it could be argued that it is also central to Ellul's critique of the triumph of technique. For example, Ellul argues that the relationship between technique and machine

> penetrates to the very core of the problem of our civilization… an inhuman atmosphere… Men now live in conditions that are less than human…Think of our *dehumanized* factories, our unsatisfied senses, our working women, our estrangement from nature. Life in such an environment has no meaning. Consider our public transportation, in which man is less important than a parcel; our hospitals, in which he is only a number (1964 [1954]: 4–5, italics added).

Dehumanization can be seen as lying at the heart of his contention that: "Men do not need to understand each other in order to carry out the most important endeavors of our times" (1964 [1954]: 132). In terms of disenchantment, he contends that technique has no place for mystery. More striking is a statement about disenchantment, which could have as easily been made by Weber: "Nothing belongs any longer to the realm of the gods and the supernatural. The individual who lives in the technical milieu knows very well that there is nothing spiritual anywhere" (1964 [1954]: 143). However, it is worth noting that Ellul argues that because people need the sacred, they make technique sacred.

More generally, just as I see McDonaldization spreading to more sectors of society and the world as a whole, Ellul has a similar view of technique. As he puts it, "The instrument tends to be applied *everywhere* it *can* be applied" (1964 [1954]: 100, italics in the original). Much of *The Technological Society*, especially in its later pages, is devoted to a discussion of the extension of technique to all sorts of sectors of society (1964 [1954]: 413). Techniques are everywhere including the state, industry, organizations, psychology (e.g. propaganda and psychotechniques), art, science, planning, biology (e.g. "human stud farms"), and sociology (managing the masses and studying public opinion). Technique not only spreads out in every conceivable direction, but it also penetrates into the "deepest recesses of the individual" (1964 [1954]: 325). More broadly, and critically, he argues that "technique *mauls* man's body and soul, we have no right to say that what is essential remains unscathed. There is, on the contrary, every evidence that what is called the 'person' is being dangerously impaired" (1964 [1954]: 393, italics added). The person is in danger because technique is everywhere and especially because it has been internalized, "it ceases to be external to man and becomes his very substance" (1964 [1954]: 6). All of this anticipates much of Foucault's (1979 [1975]: 18) thinking on the penetration of discipline into the soul of the actor.

3 The Contrast with Rationalization/McDonaldization

The biggest difference between Ellul's technique and (formal) rationalization/McDonaldization is that while the latter is restricted to means-ends rational action in formally rational organizations, there are no restrictions on technique, which appears to encompass *all* modern techniques. Thus, Weber distinguishes between formal and substantive rationality, and tends to criticize the former and value the latter more positively. I distinguish between McDonaldized and non-McDonaldized settings and focus my critique on the former. Furthermore, I view McDonaldization as a continuum and am therefore critical of those settings that are highly McDonaldized. However, there are no such differentiations in Ellul's work with the result that *all* modern techniques are the problem and they are all criticized equally. Furthermore, since every domain has a technique associated with it, everything is subject to criticism. As a result, Ellul has an even more dystopian view of the future than Weber (or me). At the beginning of the last chapter, "A Look at the Future," Ellul concludes: "We have completed our examination of the monolithic technical world that is coming to be. It is vanity to pretend that it can be checked or guided… there is 'no exit'" (1964 [1954]: 428). The only hope, if there is one, lies in the instinctive and the spiritual. No technique has emerged as yet for the spiritual, although it has in organized religion, which therefore is subjected to Ellul's critical examination.

4 Modern/Postmodern

Since he is a modernist, it is not surprising that Ellul lists rationality first, with artificiality second, in his enumeration of the basic characteristics of technique. Few things receive more attention, often critical, from modernists than rationality (for example, Simmel, like Weber, devotes a great deal of attention to it [see Turner 1986]) and few things are more criticized than artificiality (or inauthenticity versus authenticity). In its rationality, technique is critiqued by Ellul because it "excludes spontaneity and personal creativity," while artificiality "eliminates, or subordinates, the natural world" (1964 [1954]: 79). Among the other characteristics of modern technique are the "automatism" of technical choice, the tendency of technique toward self-augmentation, the monism created by the array of separate techniques, the necessary linkage of all techniques, and the universalism of techniques. It is the last five characteristics that interest and bother Ellul far more than the rationality (and artificiality) of technique. Unfortunately, it is in his discussion of these characteristics that Ellul is at his modern worst reifying the social world and offering, in postmodern terms, the grossest of totalizations and grand narratives. Now it is true that Ellul was a modernist, writing about modern techniques, and in the heyday of the modern era. However, much the same can be said of the ideas of others, notably Weber and Marx, but their work, while subject to many of the same critiques, stands up far better to them; their ideas have stood the test of time far better than Ellul's. Let us look at each of these critiques of Ellul's highly modernist work in a bit more detail, especially those offered from a postmodern perspective.

First, and perhaps most maddening, is Ellul's tendency to reify technique. For example, he contends: "Here we see the prime aspect of technical automatism. Technique itself, *ipso facto*, and without indulgence or possible discussion, selects among the means to be employed. The human being is no longer in any sense the agent of choice. Let no one say that man is the agent of technical progress" (1964 [1954]: 80). This is almost the definition, if not a caricature, of reification where a human creation – technique – is accorded a reality of its own, is endowed with the ability to act (it "does not accept the existence of rules outside itself" (1964 [1954]: 142)). Further, the human agent who created and empowers the technique is reduced to insignificance; in fact is "overpowered by technique and becomes its object" (1964 [1954]: 127). A few pages later Ellul adds another dimension – the inability of agents to prevent the choice among techniques – to this reified view of technique: "The choice between methods [in other words, technique] is no longer made according to human measure, but occurs as a mechanical process which nothing can prevent... Inside the technical circle, the choice among methods, mechanisms, organizations, and formulas [all techniques] is carried out automatically. Man is stripped of his faculty of choice..." (1964 [1954]: 82). To this, Ellul adds technique's absolute independence from, and refusal to tolerate, moral judgments (1964 [1954]: 97).

To be fair, there are places where humans are accorded some small degree of power and independence vis-à-vis technique; where technique is less reified. For example, Ellul argues that "human invention is admitted," but hastens to add that its role is "minimal" (1964 [1954]: 86). Later, this small role is accorded increasingly less significance: "the individual's role is less and less important in technical evolution" (1964 [1954]: 92). The latter is related to Ellul's grand narrative to be discussed below.

A second critique of modern social theory is its tendency to offer "totalizations," or all-encompassing theories of the world, or of significant aspects of it. Talcott Parsons is usually seen as the modern theorist who offered a single integrated theory designed to explain the social world in its entirety (see, also, the work of Niklas Luhmann). While Ellul generally does not go nearly that far, he does offer a totalizing view of the relationship between technique and modern society. The tendency toward offering totalizations is also seen in Ellul's thinking on technique per se. For example, he focuses on the integration of technique; the monism of technique; and the fact that technique includes "all of the separate techniques form a whole" (1964 [1954]: 94). However, sometimes he does go much further in arguing, for example, that: "Civilization no longer exists of itself. Every activity-intellectual, artistic, moral – is only part of technique" (1964 [1954]: 130). This seems to elevate technique to an exalted role standing above *all* modern activity and to suggest that all activity can be subsumed under the heading of technique.

Also totalizing is Ellul's tendency to think in terms of technique in general; to subsume all specific techniques under the heading of technique. Lost in this are important distinctions among techniques and even sub-techniques. Not all techniques and sub-techniques are the same or operate in the same way (for more on this, see below).

Beyond a tendency toward totalizations, Ellul is also very prone to offering a grand narrative, a single overarching history of the world. In his case, that history focuses on technique, but given his propensity for totalizations it clearly affects every nook and cranny of society. He begins with the assumption that earlier societies were "free of technique" (1964 [1954]: 65), or at least that "technique was applied only in certain narrow, limited areas" (1964 [1954]: 64). This implies a very modern tendency to value an earlier point in history and, perhaps, to imply that we would be better off returning to such a time, or at least to restructure contemporary society in such a way that technique is limited or, more radically, eliminated completely. In his day, Ellul sees himself near the endpoint of that history, "at the stage of historical evolution in which everything that is not technique is being eliminated" (1964 [1954]: 84). At the endpoint, as it eliminates "every lesser force... it will remain alone in the field" (1964 [1954]: 85). Thus, Ellul is offering a very clear and dramatic grand narrative, which begins with early societies where technique is absent or limited and ends with nothing but technique. Among other things, one is led to wonder whether, on the one hand, life is possible without technique and, on the other hand, whether a life that is nothing but technique is conceivable. It is almost impossible to accept either of the alternatives that lie on either end of Ellul's grand narrative.

These postmodern critiques bring us to what I consider the fundamental weakness in Ellul's argument. That is, all specific techniques are included under the broad heading of technique and subject to his basic critique. While Ellul discusses a number of different specific techniques, they are all combined in the category of technique and no effort is made to differentiate among them. Since all activity requires a technique, we are left with an all-encompassing critique lacking in nuance. Surely there are specific techniques that are less problematic than others; that have less in common than others and with the totality of technique? Surely there are areas of life less subject to technique than others? More, aren't there "desirable," good, techniques? Is a technique that I develop myself to serve my own needs subject to the same criticisms as techniques developed by McDonald's that lead its employees to cook hamburgers, or to serve customers at the drive-through window, in the same ways everywhere in the world that the company operates? If I develop a technique for myself for writing essays like this one, and another for books on a variety of sociological issues, those techniques work for me, and I do not seek to impose them on anyone else, are they undesirable from Ellul's perspective? If not, then it is clear that Ellul's position on technique is lacking in nuance; he needs to better distinguish among techniques.

It is in this context that my theory of McDonaldization is most at variance with Ellul's theory of technique. While I, too, operate as a modernist, and can be accused of offering a grand narrative, a totalization, and even a reified view of the process McDonaldization, my view is far more limited in scope than Ellul's. As indicated above, I think in terms of degrees of McDonaldization, as well as of McDonaldized and non-McDonaldized settings. This is clear throughout my work on the topic, but is perhaps clearest in my essay on McDonaldization in which instead of thinking in terms of an "iron cage," I operate with the view that McDonaldized systems should be seen as "islands of the living dead" (Ritzer 2003). This indicates two crucial differences with Ellul's theory. First, the fact that McDonaldized systems are only a series of islands is meant to indicate, as is the case in Foucault's (1979 [1975]) "carceral archipelago," that the sea, or spaces, between those islands are less – or even non –, McDonaldized. Ellul's theory offers no such spaces in the contemporary world free of modern techniques. Second, while the critique that these islands are "dead" is consistent with Ellul's contention that technique is lifeless, the idea that there is "life" on these McDonaldized islands is meant to indicate that there are positive aspects of McDonaldization and that people really enjoy their time on those islands. Therefore while Ellul is uniformly critical of technique, I see a more positive side to McDonaldization. In fact, as I have often pointed out, the basic characteristics of McDonaldization – efficiency, calculability, predictability, and control – can all be seen as desirable, at least in moderation. The critical focus in the McDonaldization thesis is on the irrationalities spawned by the process.

5 Prosumption and Technique

Like most modernists, Ellul (1964 [1954]: 65) tends to differentiate between production and consumption (Ritzer and Jurgenson 2010), as well as the techniques associated with them. However, in my view (Ritzer 2010), society has always been dominated by prosumption, or the reality that production always involves consumption (of, for example, raw materials, labor, time and power) and that consumption always involves production (for example, of what is consumed, or of the experience of consumption). Ellul's theory leads to the assumption, then, that as with everything else, technique encompasses prosumption, or that there is, must be, a technique for prosumption. In my work on McDonaldization I discuss the ways in which McDonald's (and other fast food restaurants and rationalized businesses of various sorts) puts its customers to work. For example, McDonald's makes it clear that they are to clean up after themselves rather than having paid employees do the work. McDonald's even offers tools – for example, clearly labeled garbage receptacles – to help customers understand the technique they are to use. I even remember my children "training" me in the steps involved in cleaning up after a meal at McDonald's. This "employment" of consumers is occurring in more and more sectors of society with, for example, airlines, even with ever-higher airfares, increasingly asking passengers to load their own baggage (as carry ons), help clean up the plane, and instructing them in what to collect, before it lands so that there is less, or no, work for employees. Both of these examples illustrate that there are techniques for prosumption, although interestingly such techniques cannot be imposed on paying customers in the same way that they are on paid employees. Much greater subtlety is needed in order to get customers to use the desired techniques than is employed in the case of employees. This suggests yet another nuance needed for a deeper understanding of technique.

While there are certainly techniques for prosumption of various types and forms, the fact is that the concept itself builds on the whole idea of agents who produce consumption as they consume production. Because the prosumer is actively involved in production and consumption, there is a lesser likelihood of being controlled by the techniques of prosumption. Of course, whether this turns out to be the case will require the test of time. Various entities (like fast food restaurants and airlines) will endeavor to create techniques of prosumption and to lead prosumers to use them. In fact, the more we serve as prosumers, and grow increasingly accustomed to discerning and following the techniques associated with prosumption, the more likely we will be to accept and follow them in other contexts. While it may be that prosumers will have greater ability to resist the control exercised by these techniques, the history of their imposition on producers and consumers at least gives us pause in over-romanticizing the capacities of the prosumer as agent.

The central domain in which to think this through is Web 2.0 and its websites which are based on the centrality of the prosumer. It is in those sites that we need to examine and think about the dialectic between the techniques that exist there and are to be used by prosumers and the ability of the latter, as agents, to evade them or, more importantly, to create techniques of their own on those sites and more generally on the Web. Fundamentally, in Ellul's terms, there are techniques in place for prosuming on such Web 2.0 sites as Wikipedia, Facebook, and Twitter. For example, there are methods for writing, or adding content to, a Wikipedia entry and if those methods are not followed, the added content can be deleted by the administrators of the site. Other users can modify or delete content that they think has been added improperly. Similarly, the Facebook wall and the rules that define and inform it place limits on what can and cannot be done on it. Most extremely, the 140 character limit on Twitter's tweets places a powerful constraint on those who use the instant-messaging system.

In fact, one needs to learn the proper technique for using Web 2.0 sites. For example, writing meaningfully in 140 characters is quite unnatural for most people, at least until they learn the technique. To take another example, writing a blog requires a different technique than writing almost anything else. For example, it is best to make the main point at the beginning of a blog, rather than building toward it gradually, since readers of blogs are likely to grow impatient quickly and to not read beyond the first few sentences.

However, while users of these Web 2.0 systems are limited in various ways, and many others, there is also considerable leeway in terms of what users (agents) can do and say on those sites. Of course no technique is ever totally constraining, but techniques are far less constraining on Web 2.0 sites that are designed to have content produced by the consumers of those sites.

While that is the situation today, it is clear that there is an ongoing struggle between those who create the techniques that define these sites and their prosumers. More importantly, what is lost sight of is the fact that prosumers – like producers and consumers, although they are less able to do so – develop their own techniques for using Web 2.0 sites and especially for adding content to them. This points, again, to perhaps the greatest weakness in Ellul's approach to technique. Because he focuses on the techniques created by systems, and because he lacks a sense of agency at least in relationship to technique, he fails to recognize that agents often create techniques that are at least in part their own, even in systems with powerful and constraining technical systems. It is at least as important that we focus on those agential techniques as well as on the fact that the interaction of such techniques with those that are generated by systems often end up creating unique techniques that are at variance with those envisioned and generated by systems. It may be that the very nature of prosumers, and their position at the intersection of production and consumption, makes them better able to create such unique techniques. However, returning to the more pessimistic perspective of Ellul and Weber, and to one that I am more comfortable with, the systems involved, on and off Web 2.0, will continue to seek to codify and to control the techniques developed by prosumers. They are also likely to create and impose new techniques, as well as develop and refine sites that lead prosumers to adopt techniques that the sites prefer they use.

It is worth noting that it is increasingly difficult to deal with online phenomena as if they are disconnected from offline phenomena. The two are converging in various ways. Among other things, this means that the resolution of the struggle between system – and user – generated techniques online is likely to affect the relationship between offline systems and users (and vice-versa). While there are trends toward prosumption in both domains, the prosumer is stronger online, especially on Web 2.0, than offline. Thus, it is online prosumers who offer the greatest hope in the conflict between system- and user-generated techniques.

6 Conclusion

It is clear that Ellul was on to something with his focus on technique, but the latter is a far more complex and contested domain than he imagined. We need a more refined and differentiated sense of technique, as well as one that is better adapted to changes over time rather than simply assuming unidirectional and continuous historical development. Above all, we need a less reified view of technique, one that accords the human agent a central role in creating, operating, and especially contesting it. Without such changes in perspective, we have no way of truly understanding the role of technique in the social world, including the new worlds emerging on Web 2.0 (and beyond). All of those worlds are being created by humans, and within them there are vigorous and fateful conflicts between agents who "produce" the systems and those who "consume" them. However, from the point of view of this paper, we know that both are "prosumers" with the result that this is a conflict taking place *between* prosumers and *within* the world of prosumption. Such a viewpoint not only gives us a more accurate sense of the contested field that is the social world and Web 2.0, but also prevents us from adopting the kind of reified perspective that is so damaging to Ellul's views on this domain.

References

Bauman, Zygmunt. 1991. *Modernity and ambivalence*. Ithaca: Cornell University Press.
Bauman, Zygmunt. 1992. *Intimations of postmodernity*. London: Routledge.
Bryman, Alan. 2004. *The disneyization of society*. London: Sage.
Drane, John. 2008 [2001]. *The McDonaldization of the church*. London: Darton, Longman, and Todd.

Ellul, Jacques. 1964 [1954]. *The technological society*. New York: Vintage Books.
Foucault, Michel. 1979 [1975]. *Discipline and punish: The birth of the prison*. New York: Vintage.
Hayes, Dennis, and Robin Wynyard (eds.). 2002. *The McDonaldization of higher education*. Westport: Bergin and Garvey.
Hayim, Gila. 1978. Modern reality strategies: An analysis of Weber, Freud and Ellul. *Human Studies* 1: 315–329.
Izberk-Bilgin, Elif, and Aaron Ahuvia. 2011. eBayization. In *McDonaldization: The reader*, 3rd ed, ed. George Ritzer, 425–432. Thousand Oaks: Pine Forge Press.
Kalberg, Stephen. 1980. Max Weber's types of rationality: Cornerstones for the analysis of rationalization processes in history. *The American Journal of Sociology* 85: 1145–1179.
Lyotard, Jean-Francois. 1984 [1979]. *The postmodern condition*. Minneapolis: University of Minnesota Press.
Maley, Terry. 2004. Max Weber and the iron cage of technology. *Bulletin of Science Technology Society* 24(1): 69–86.
Mannheim, Karl. 1936 [1929]. *Ideology and Utopia*. New York: Harcourt, Brace and World.
Mannheim, Karl. 1940 [1935]. *Man and society in an age of reconstruction*. New York: Harcourt, Brace and World.
Menninger, David. 1981. Politics or technique: A defense of Jacques Ellul. *Polity* 14(1): 110–127.
Mitzman, Arthur. 1971 [1969]. *The iron cage: An historical interpretation of Max Weber*. New York: Grosset and Dunlap.
O'Byrne, Daniel J., and Alexander Hensby. 2011. *Theorizing global studies*. New York: Palgrave Macmillan.
Ram, Uri. 2004. Glocommodification: How the global consumes the local McDonald's in Israel. *Current Sociology* 52: 11–31.
Ritzer, George. 1975. Professionalization, bureaucratization, and rationalization: The views of Max Weber. *Social Forces* 53(4): 627–634.
Ritzer, George. 1997. *Postmodern social theory*. New York: McGraw-Hill.
Ritzer, George. 2003. Islands of the living dead: The social geography of McDonaldization. *American Behavioral Scientist* (special issue) 47(2): 119–136.
Ritzer, George. 2010. Focusing on the prosumer: On correcting an error in the history of social theory. In *Prosumer revisited*, ed. Blattel-Mink Birgit and Hellman Kai-Uwe, 61–79. Wiesbaden: VS Verlag.
Ritzer, George. 2011. *The McDonaldization of society*. Thousand Oaks: Pine Forge Press.
Ritzer, George, and Nathan Jurgenson. 2010. Production, consumption, prosumption: The nature of capitalism in the age of the digital "prosumer". *Journal of Consumer Culture* 10(1): 13–36.
Turner, Bryan. 1986. Simmel, rationalization and the sociology of money. *The Sociological Review* 34: 93–114.
Watson, James L. (ed.). 2006 [1997]. *Golden arches east: McDonald's in east Asia*. Stanford: Stanford University Press.

Chapter 4
Are We Still Pursuing Efficiency? Interpreting Jacques Ellul's Efficiency Principle

Wha-Chul Son

Efficiency is one of the most significant elements that characterizes modern technology.[1] However, it is a notion that philosophers of technology do not pay much attention to. Jacques Ellul was no exception to this charge and, although he emphasized its concept repeatedly, he did not elaborate the concept of efficiency. It is, thus, important to review Ellul's theory in terms of the so-called "efficiency principle." This allows us not only to appreciate Ellul's philosophy, but also to continue developing his insights.

The following discussion is in five sections. After surveying the efficiency principle as presented in Ellul's works, three critiques of the principle are introduced and then followed by a new interpretation defending Ellul from those critiques. A further elaboration of the efficiency principle is then presented. Finally, an application of Ellul's insight relevant to our era is given.

1 The Efficiency Principle in Ellul's Philosophy of Technology

Although Ellul himself did not use the term "efficiency principle," it effectively describes his claim that efficiency is the only criterion in all the decision making processes in technological society. Several passages reflect this principle:

> The intervention of rational judgment in the technical operation has important consequences… Reason… takes account of the fixed end of technique – efficiency. It notes what

[1] At the outset, the terms "technology," "technique" and "techno-logy" should be clarified. In this paper, I use "techno-logy" to emphasize *technologie* in terms of "discourse on technique" (Vanderburg 2004 [1981]: 26–27; Ellul 1990 [1988]: xvi). Thus, I use "techno-logical bluff" rather than "technological bluff," except when referring to *The Technological Bluff* as a book title. "Technology" is used in a more loose, general, neutral, and conventional way. "Technique" is seldom used except when Ellul's works or works on Ellul are referred to (cf. Son 2005b: 232).

W.-C. Son (✉)
Global Leadership School, Handong Global University, Pohang, South Korea
e-mail: phtech@handong.edu; phtechson@gmail.com

H.M. Jerónimo et al. (eds.), *Jacques Ellul and the Technological Society in the 21st Century*, Philosophy of Engineering and Technology 13,
DOI 10.1007/978-94-007-6658-7_4, © Springer Science+Business Media Dordrecht 2013

> every means devised is capable of accomplishing and selects from the various means at its disposal with a view to securing the ones that are the most efficient, the best adapted to the desired end. Thus the multiplicity of means is reduced to one: the most efficient… [I]n every field men seek to find the most efficient method (Ellul 1964 [1954]: 20–21).
>
> Technical progress today is no longer conditioned by anything other than its own calculus of efficiency (Ellul 1964 [1954]: 74).
>
> In reality, to-day what justifies the means is the means itself, for in our day everything that 'succeeds,' everything that is effective, everything in itself 'efficient,' is justified (Ellul 1967 [1948]: 70).

The pursuit of efficiency is a recent phenomenon characterizing our contemporary society. "When studying the old techniques, one is extremely surprised to see how unimportant efficiency was as a decisive or determining notion" (Vanderburg 2004 [1981]: 29). Ellul viewed this change as fundamental and argued that when efficiency became the sole criteria for technical choice, "it was felt that not only the traditional but the deepest instinct of humankind had been violated" (Ellul 1964 [1954]: 73).

The famous "characterology of technique" presented in *The Technological Society* (Ellul 1964 [1954]: 79–147) cannot be explained without the efficiency principle. Efficiency is the only criterion for technical choice ("automatic choice in technique"). The driving force for the "self-augmentation of technique," such as the many alterations which stemmed from the steam engine or the "necessary linking of techniques" in examples like the connections between commercial technique and transportation technique, is the efficiency principle. "Monism," i.e. technique cannot be separated from its use, indicates that moral judgments of the consequences of technique are ignored while efficiency is highlighted. "Universalism" reiterates that the principle of efficiency is applied to all areas of human life, geographically and quantitatively.

Furthermore, the concept of autonomous technology, which is often considered the trademark of Ellul's philosophy (Son 2005a), can be reiterated in terms of the efficiency principle. Ellul argues that "the complete separation of the goal from the mechanism, the limitation of the problem to the means, and the refusal to interfere in any way with efficiency" (1964 [1954]: 133) are the basis of autonomous technology. This contradicts the aforementioned "deepest instinct of humankind," namely, the freedom of human being. Langdon Winner explained the concept of autonomous technology as "the question of human autonomy held up to a different light" (1977: 43). The violation of human autonomy begins when efficiency becomes the basis of all judgment.

Alexander (2008) proves by historical study the significance of efficiency in modern society[2] and shows that the pursuit of efficiency is closely linked to the modern ideal of total human control over the preservation of desired conditions as well as the process of dynamic change in different spheres. She analyzes various usages of this concept throughout history with examples that indicate that it still

[2] I thank Carl Mitcham for directing my attention to this work.

commands a strong presence even in the postmodern era. While not sharing Ellul's strong normative (or negative) position on the issue of efficiency, Alexander justifies and supports Ellul's concern for efficiency in technological society.

2 Critiques of Ellul's Efficiency Principle

Unfortunately, the efficiency principle is not as intuitively acceptable as the concept of autonomous technology. At least to those who are sympathetic to critics of modern technological society, it is relatively easy to understand that the development of technology is not under the full control of human beings. However, it is unclear whether the pursuit of efficiency is the sole reason for its uncontrollability. What does Ellul mean by "efficiency"? Is it true that everything is determined by efficiency in technological society? Does he view efficiency as negative? In the process of analyzing Ellul's theory as a whole, these questions deserve our attention.

2.1 Efficiency for What?

The foremost criticism of Ellul's principle of efficiency is the fact that the term "efficiency" is ambiguous. Because the concept of efficiency is context-dependent (Mitcham 1994: 225), using the term without referring to its context only leads to uncertainty. Efficiency is always the efficiency of or for something. "[A] discussion of the concept of 'efficiency' makes sense only against the background of a social institution" (Carpenter 1983: 69).[3] Depending on time span, spatial situation, economic condition, or available technologies, all of which Carpenter defines as "social institutions," efficiency could be measured differently. Economic efficiency and technological efficiency, for example, are not necessarily the same.

Ellul's claim that efficiency is the sole criterion for judgment in all areas of technological society is too general. When no particularities are noted, such a claim is over-inclusive and inaccurate, explaining virtually everything or nothing. This observation then leads to denouncing Ellul's severe critique of modern technological society.

2.2 Are We Still Pursuing Efficiency?

While the first critique of the efficiency principle is conceptual, the second concerns historical validity, namely, whether all technological decisions in history have been made according to efficiency criteria. One can easily think of other elements, such

[3]Referring to the fact that efficiency can be calculated in terms of one specific result, Ellul himself made a similar comment: "For true efficiency, not only must the rational aspect of the machine be taken into account, but also its adaptation to the environment" (Ellul 1964 [1954]: 75).

as cultural, political, and aesthetic considerations, interfering with technological judgments. The importance of design and color in electronic products, for example, or environmentally friendly designs indicate that efficiency is not everything.

Social constructivists have argued that the history of technology does not confirm the efficiency principle. They point out that technological development is the outcome of complicated interactions among relevant social groups. Each group has a different interpretation of a given artifact, which does not have much to do with efficiency. Pinch and Bijker's (1987) example of the bicycle clearly shows that current bicycle design was chosen by chance. It was not because of efficiency that the design of bicycle with a huge front wheel failed to become the standard design. It was the tacit triumph of the "relevant social group" which chose safety rather than speed.

The main target of this approach was technological determinism, which refers to the idea that technological development follows a fixed path and that technology unilaterally influences society. Ellul is one of the representative theorists they criticize. According to social constructivists, the autonomy of technology is misleading, because the history of technology indicates otherwise.

This critique is succinctly expressed by Andrew Feenberg. Describing ten paradoxes of modern technology, he argues "efficiency does not explain success; success explains efficiency" and calls this "the paradox of the frame" (2010: 7). He claims that technologies survive not because of their efficiency but because of "the contingent circumstances of success and failure" in history. According to this view, the efficiency of existing technologies has been "constructed" in order to justify their choice. As a result, the efficiency principle can be viewed as ahistorical, oversimplified, and exaggerated.

2.3 *The Possibility of Modifying the Efficiency Principle*

Based on the aforementioned context-dependent characteristic of efficiency, one might argue that the main problem of technological society is not that efficiency is stressed but that the context or goal of pursuing efficiency is not clearly set. Once the context is clarified, the efficiency principle can be the starting point of an alternative progress in technological society. If the final goal of a practice or an artifact is good, then why not try to find an efficient way to realize it? For example, there is nothing wrong with seeking the most efficient method of manufacturing an environmentally friendly car. Therefore, the primary issue is not to stop seeking efficiency, but rather to establish appropriate goals of technological development.

This position does not contradict Ellul's critique of technological society, but rather suggests possibilities of redirecting the efficiency principle. Willem H. Vanderburg's notion of the "ecology of technology" deals with this point:

> The economy of technology strips away the contexts of human life, society, and the biosphere, leaving only the inputs and outputs that connect these contexts to the technology. The ecology of technology includes the consideration of undesired outputs, and the meaning

> and value of all inputs and outputs by means of which technology is embedded in, depends on, and interacts with its contexts, in so far as this is relevant to the particular issue at hand (Vanderburg 2000: 16).

While the "economy of technology" in technological society represents Ellul's efficiency principle, Vanderburg's suggestion of preventive engineering calculates efficiency in a longer and broader perspective.

Benello argues: "We cannot accept Ellul's claim that technique is coterminous with efficiency except in the narrowest and most mechanistic sense" and "Ellul has too easily accepted the rhetoric of those who uphold the system of instrumental rationality" (1981: 104–105), although the social system based on instrumental rationality is not efficient at all due to all the social costs caused by the mechanical domination of society. It is important to recognize this inefficiency, because the contradictions in instrumental rationality could bring about dialectical change. Benello criticizes Ellul for not allowing room for the possibility of dialectical revival of humanism in technological society (1981: 104–105). He agrees with Carpenter who suggests that "we could more appropriately fashion our frames of reference differently" (1983: 75).

This position proposes a modest solution that accommodates the positive aspect of efficiency as well as Ellul's critique of technological society. On the one hand, it complements the lack of concrete alternatives in Ellul's theory on technology. On the other hand, however, his claim that the efficiency principle is the key problem of technological society is made obsolete.

3 A New Interpretation of the Efficiency Principle: "In the Name of" Efficiency

How shall we understand Ellul's notion of efficiency? Is there any way to interpret the efficiency principle other than as the result of a vague and inaccurate observation of technological society or as an unfinished concept that can be supplemented by further elaboration?

In "Reading Jacques Ellul's *The Technological Bluff* in context," I suggest a defense of Ellul's diagnosis of the efficiency principle. I argue that the principle should be understood as describing a situation in which everything is justified *in the name of* efficiency. In other words, in technological society, "all techniques are supposed to be, and justified as, the pursuit of efficiency... [T]echnological development is not pursued in order to achieve more efficiency, but *in the name of* more efficiency" (Son 2004: 525, italics in the original).

As it is, this interpretation serves to provide not only good answers to the questions concerning the principle, but also a consistent understanding of Ellul's position. The latter is important in the sense that an advanced Ellulian prescription for the future of technological society can be developed without rejecting his diagnosis. The following is devoted to defending Ellul concerning three criticism raised in Sect. 2.

3.1 *Efficiency for Justification*

The vagueness of the term "efficiency" is a double-edged sword. On the one hand, it is useful for criticizing Ellul's claim that technological society is run by the efficiency principle. On the other hand, it is responsible for the blind acceptance and justification of current technological development.

According to my interpretation, the efficiency principle does not, of necessity, refer to efficiency in terms of input and output. It describes a situation in which people accept any device or activity once it has been qualified as "efficient." The interval between input and output, the fulfillment of the final goal, as well as judgment criteria can be adjusted or renamed in order to describe any technological activity or artifacts as "efficient." Therefore, what deserves our attention is the possibility of convenient adjustments in the calculation of efficiency. Carpenter states this clearly:

> [There is] word magic associated with using the term 'efficiency.' We admit, on reflection, that every calculation of efficiency involves the drawing of boundaries, a deliberate focusing on some properties and the ignoring of others. Yet, when the conventions for bounding our calculations are widely accepted to the extent of being institutional norm as themselves – when we are given institutional license for thinking in certain habitual ways – then limited measures of 'efficiency' become approved usage (Carpenter 1983: 70).

According to this interpretation of the efficiency principle, the notion of efficiency is subject to "word magic" precisely because of its vagueness. While the notion of efficiency requires the recognition of the final outcome and possibility of calculation (Cózar 2000: 92), often-stated descriptions of technological efficiency are far from sufficient to calculate such efficiency. If a smart phone is efficient, what does that mean? Should we include the time for downloading applications for chatting online in the calculation of its efficiency, or should we only consider the possibility that e-mail can be checked anywhere, anytime? Depending on which elements are included in the calculation, anything can be considered efficient. In such a rationalization, a nuclear power plant and a regional energy system based on solar energy can both be considered efficient.

Commenting on how the economy works in a technological society, Ellul pointed out that there was no criterion for the measurement of "efficiency":

> It is efficiency and success that lead history to adopt a certain direction – not man who in some sense makes a decision. The problem does not concern personal decision or preference; it is a question of discerning what seems most probable. At the present moment, what system is most efficient? I insist on the phrase *at the present moment*. It means nothing to explain that liberal capitalism was extraordinarily efficient a century ago (Ellul 1964 [1954]: 183–184, italics in the original).

This elastic (uncertain) characteristic of the efficiency principle strengthens rather than weakens Ellul's position. The problem is not that Ellul used the term inaccurately, but that an inaccurate concept was being used for justifying and perpetuating the current trend of technological society.

3.2 *Overcoming the Accusation of Determinism*

According to this orientation, the efficiency principle no longer implies a linear technological development through a fixed path. The accusation of determinism, as promoted by social constructivists, is not so damaging, since the efficient principle no longer focuses on the efficiency achieved. Historical evidence for the principle is no longer based on whether technological development was efficient in terms of actual input and output, but whether efficiency was considered the most important reason for technological changes.

Feenberg's paradox of the frame (2010) can easily accommodate this interpretation. What he described as "paradox" is exactly what Ellul viewed as the problem of modern technological society. On the one hand, the pursuit of efficiency characterizes the success; on the other hand, everything successful is reiterated as efficient. However, this "frame" of success or efficiency does not leave room for human freedom. Whether it is efficient or not, every technological choice should be justified in terms of efficiency. The influence of society on technology need not to be denied in order to point out the nonfreedom in technological society (cf. Ellul 1990 [1988]: 411), since every technology has to be justified in terms of efficiency.

Is the diagnosis of the efficiency principle still applicable to contemporary society? It seems that there has not been much change since Ellul began his analysis of technological society. Examples of the efficiency principle are everywhere. Alexander (2008) provides good evidence that the notion of efficiency is widely used in the areas not only of technology, but of law, economy, education, and even biology. That efficiency has been considered as a moral value was shown in the heated dispute over *Time on the Cross* (Fogel and Engerman 1995 [1974]) which highlighted the efficiency of slavery[4] (Alexander 2008: ch. 6). Even the widespread trend of postmodernism seems to have had little impact on the pursuit of efficiency. In technological society, efficiency makes the best explanation for everything.

3.3 *What Is the Output?*

It might be argued that my interpretation rescues Ellul from the accusation of determinism. However, that does not exclude the possibility of clarifying the "final goal" in order to redirect the efficient principle. If this principle is used for justification, why not clearly set the goal so that it can overcome the problems of technological society? This interpretation is compatible with the concept of efficiency which remains neutral, meaning that it can be applied and evaluated positively or negatively depending on the final cause.

[4]Later, Fogel wrote another book, *Without Consent or Contract* (Fogel 1994 [1989]), in which he changed his original position. Alexander (2008) uses this example in order to show the confusion caused by different understandings of efficiency.

However, the observation of technological development being justified in the name of efficiency rightly accommodates the fact that it is impossible to calculate the efficiency of a given technology, due to the "unpredictability" of its long-term effects (Ellul 1990 [1988]: 77ff). Since "man can never foresee the totality of consequences of a given technical action" (Ellul 1964 [1954]: 105), it is utterly impossible to define what the final outcome of a certain technology is and what elements should be included in its calculation. The unexpected, long-term, and non-mechanical consequences of modern technology are beyond human imagination. Ellul rejects the suggestions for a new calculation of efficiency, because they presuppose the possibility of measuring the immeasurable.

4 The Categorical Refusal of Efficiency

Ellul's diagnosis of technological society in terms of the efficiency principle has been proven to be defendable. Apart from defending his theory from critiques, however, we need to formulate Ellul's own position concerning efficiency and the efficiency principle. In other words, the remaining task is to clarify how the notion of efficiency should be situated in Ellul's philosophy of technology. This will lead to the development of an Ellulian prescription for the current situation on three levels, which roughly parallel the problems regarding the efficiency principle: conceptual clarity, historical evidence, and the possibility of modification.

4.1 Efficiency as a Tool of Mastery

After dealing with efficiency throughout history, Alexander rightly observes that efficiency functions not only as a measurement for tools but also as a tool itself:

> Efficiency, however, even in its technical form, was a tool of control and not a mere technical measurement, disciplinary at its inception and increasingly political after it had reached conceptual maturity. As a measurement, it has an apparently objective form, but its history is as a tool designed to make the natural and human worlds conform to the way in which they are intellectually understood (Alexander 2008: 169).

This remark reveals what lies hidden behind the notion of efficiency, namely the desire to control and the mastery of everything. The "quintessentially modern… faith in the powers of the human intellect to comprehend the world and the belief that the world itself did indeed correspond to the categories and methods of human comprehension" (Alexander 2008: 165) is the basis for the notion of efficiency. Its prevalence in modern society goes together with the assumption that it is possible to control all elements including humans, to plan all processes, and to measure everything involved. As Alexander observes, the notion of efficiency is closely linked to "the modern idea of humankind as the author of its own fate," but one should note that humanity cannot avoid becoming the object of efficient manipulation.

Those assumptions are precisely what Ellul found abominable. Not only are they not true, but also they disguise the reality of technological society and lead to a situation of nonfreedom. In Ellul's theory, the pursuit of efficiency in modern technological society needs to be eradicated in order for an alternative to the current society to be established.

4.2 The Efficiency Principle as a Techno-Logical Bluff

The efficiency principle successfully bridges *The Technological Society* (1964 [1954]) and *The Technological Bluff* (1990 [1988]) as it penetrates the different stages of modern technological society that Ellul attempted to describe in his technology trilogy.

In *The Technological Society*, Ellul focused on the characteristics of modern technique. He explained the autonomy of technology as humanity's inability to refuse to calculate efficiency in terms of income and outcome. If a machine could produce more than another method, then all else being equal, humanity has no authority to choose the latter. In *The Technological System* (Ellul 1980 [1977]), Ellul describes the situation in which humanity is not only deprived of its autonomy but also becomes a part of the technological system.

In *The Technological Bluff* (1990 [1988]), Ellul focused attention on technology as the discourse of technique rather than on technique itself. The development of modern technique leads to a situation in which technique is so taken for granted that people feel comfortable and completely at ease with their nonfreedom. Various types of techno-logical bluffs fascinate people so that all kinds of absurdities which naturally follow from the technological system are accepted without resistance.

The efficiency principle can thus be seen as the prototypical techno-logical bluff. In it, the notion of efficiency justifies all technological development. Different gadgets and activities are explicated as efficient without clear evidence or support. People are ready to believe it, not bothering to calculate the actually efficiency, even when it is possible. The autonomy of technology is thus complete. Eventually, the efficiency principle could become obsolete, as the completion of the techno-logical bluff would result in the abolition of the need for any justification for technological development (Son 2004: 526).

The concept of techno-logical bluff indicates clearly how the notion of efficiency has evolved through the different stages of technological society. The efficiency principle is not a simple description of what is going on in technological society, but a mirror reflecting the core problem of our age.

4.3 Reverse Adaptation

The efficiency principle points directly to the fact that the modern technological development has no end (*telos*) (cf. Ellul 1980 [1977]: ch.11).[5] It refers to the situation in which everybody is marching on in the name of efficiency, while nobody knows the destination. The problem is not that its goal is bad or unclear, but that the goal itself has become redundant. The efficiency itself, not its goal, has become the prime mover of technological development.

When a goal for a technological innovation is suggested, it is either faked or redefined in terms of efficiency so that the current path of autonomous technology continues. Winner (1977: 238–251) discusses the latter case calling it "reverse adaptation" referring to the case in which the end is readjusted to fit the means. Winner's critical analysis of the appropriate technology movement in America in the 1970s (1986: ch. 4) and risk assessment (1986: ch. 8) clearly show how modern technology obscures any attempt to redirect its development path. Appropriate technology is suggested as an alternative to replace the current technological trajectory, that is, to establish a new system that would be environmentally friendly and sustainable. However, the boom of appropriate technology turned out to be another popular trend that compliments the existing order:

> Rather than attempt to change the structures that vexed them, young Americans growing older were settling for exquisite palliatives. If the 1960s proclaimed, 'Let's see if we can change this society,' the 1970s answered, 'Let's get out of this skyscraper and go jogging!' (Winner 1986: 76)

As a result of this attitude, the existing order, dependent upon the efficiency principle, was not even touched. Appropriate technologies only served to maintain a society that they claimed to overcome.

Being introduced as a method to deal with the uncertainty of modern technology, risk assessment has been used to support the *status quo* of existing technological development. In other words, assessing risk leads to the desire to reduce risk, rather than to fundamental reconsideration of its source, thus affirming what is at stake, whether it may be a nuclear power plant or human cloning. Winner argues that the risk analysis created a new kind of "conservatism" (1986: 139). Risk is changed into a problem to be solved thus inviting the measurement of efficiency.

The possibility of reverse adaptation indicates that the chance to set a new goal to confront technological society is very slim, because it is soon restated in terms of the logical of problem solving and then becomes an efficiency issue. In this, Winner's politics of technology shares the view with Ellul that the true alternative for technological society is not compatible with efficiency.

[5]Hans Jonas described this as the "dialectics or circularity of means and ends" (1979: 35).

5 Overcoming the Efficiency Principle

5.1 From the Efficiency Principle to an Ellulian Solution

Since the efficiency principle represents the fundamental problem of technological society, it could be a good starting point for anticipating an Ellulian "alternative" to technological society. However, Ellul does not give many hints about the future of technological society, remaining faithful to his "realism" rather than giving false hope. Nevertheless, the examination of his efficiency principle naturally leads us to seek an alternative to replace it, at least in the realm of technology. What kind of technology do we want to have? Which direction should technological development now take for a better future? Is it possible to have such a technology neither trapped by the efficiency principle nor suffering from reverse adaptation?

An answer can be found in Ellul's description of traditional technologies. In ancient Rome technology "was directed toward a precise end: the internal coherence of society. This technique was not self-justifying, it did not have as its *raison d'être* its own self-development and it was not imposed from the outside" (Ellul 1964 [1954]: 31).

The 1968 Movement and the increasing use of the computer provided another hint for such a hope (Vanderburg 2004 [1981]: 45). They aimed at a different kind of progress, which could be achieved without referring to efficiency. At least in the beginning, the human purpose was highlighted and the efficiency to achieve those goals was not the main issue.

Based on these observations, I submit that the opposite of Ellul's "*the totality of methods rationally arrived at and having absolute efficiency*" (Ellul 1964 [1954]: xxv, italic in the original), is actually a "purpose driven technology."[6]

5.2 Purpose Driven Technology

Purpose driven technology is a technological innovation which finds its justification in the purpose of the innovation, but not in the name of efficiency. It is a general principle of technological practice. This is well described by Cózar when he refers to Micham's comparison of the naturalistic fallacy and the efficiency fallacy. Just as "being natural" does not imply "being good," the achievement of efficiency is still open to the question "but is it good?" "Therefore, one should define the goals one judges as good and only then, if appropriate, look for the means to achieve them efficiently" (Cózar 2005: 603). The development of a technology, therefore,

[6]I borrowed this term from the famous book by *The Purpose Driven Life* (Warren 2004). I do not wish to convey a religious connotation for my proposal here, but the central position that God's purpose occupies in that book should be paralleled with the importance of human agent's purpose in my suggestion.

should begin with a justification of its purpose. The criterion of efficiency, in a space-and-time-specific sense, could be used for minor and concrete decisions but not for the justification of a given project.

The main value of purpose-driven technological innovation, then, is that it revives humanity's initiative in the realm of technology. While the efficiency principle highlights the autonomy of technology, purpose driven technology emphasizes human control.

5.3 *Examples of the Purpose Driven Technology*

The purpose of technology could be considered on different levels, from general to specific. One could search for a general purpose of technological innovation as a whole, while others focus on a particular technology. It could refer to technology as an action, but also to technology as a product.

An example of the general purpose for the contemporary technological society could be a new paradigm of technological development that will eventually overcome the technological divide. The technological divide has been considered an undesirable byproduct of technological development. Various attempts have been made to promote technological innovation in underdeveloped countries, but they have been largely unsuccessful. However, once the overcoming of technological divide is seen as the purpose of technological development, such development will be shaped in a completely different way.

There exists a specific and successful example of a purpose driven technology. Robinson's "energy backcasting" (Robinson 1982) is an alternative to the common forecasting of energy consumption in energy policy making. This means determining the target consumption level at a certain moment in the future and then establishing a relevant energy policy. While energy forecasting considers future energy consumption as something already given and focuses on how to supply it, the backcasting method emphasizes the agency in energy demand and supply.

Purpose driven technology could be realized in the engineering design process. Sclove's "design criteria for democratic technology" is a good example of this (1995: 98). He argues that the norm of democracy should be reflected in technology as early as at the stage of the design process. Neil Postman further explains how the purpose of technology can become the paramount concern of technological innovation:

> The most obvious question to be asked about any new technology… is, *What is the problem to which this technology is the solution?* This question needs to be asked because there are technologies that are employed – indeed, invented – to solve problems that no normal person would regard as significant (Postman 1999: 42).

Various products of the recent appropriate technology movement targeted for the underdeveloped world can thus be seen as purpose driven technology (Smith 2007). Such products not only reflect the immediate needs of local people but also their cultural, economical, and natural environment. These products have not surrendered to the reverse adaptation, since the purpose of technology supercedes the efficiency.

The common feature of these examples indicates that efficiency is being ignored at a minimum level. Since they highlight the purpose rather than the efficiency of the given technologies, it is unlikely that the purpose driven technology will fall victim to techno-logical bluff. The danger of reverse adaptation still remains since it is difficult to imagine that technology will ever rid itself of the notion of efficiency. To place efficiency appropriately within the realm of technological enterprise is another task for philosophers of technology.

6 Conclusion

In a passage that could easily be used to accuse him of pessimism, Ellul argues:

> Following Hegel, Marx, and Kierkegaard, I have often said that we show our freedom by recognizing our nonfreedom... the only thing we can do is set them at a crucial distance, for it is by being able to criticize that we show our freedom. This is the only freedom that we still have if we have at least the courage to grasp it. Nothing is more certain (1990 [1988]: 411).

According to Ellul, the efficiency principle is, thus, the primary source of nonfreedom in technological society. Any step toward freedom can be taken only after close analysis of the problem confronted in technological society. The current investigation is in accord with Ellul's recommendation in the sense that it begins with a close examination of the efficiency principle as a key to understanding his analysis of technological society and then offers an Ellulian alternative. The concept of purpose driven technology not only reflects Ellul's insight and zeal for freedom, but also points to the hope hidden in his theory. The fulfillment of the hope is, of course, not as certain as the freedom of nonfreedom.

Acknowledgment I thank Gayle A. Henrotte, my colleague at Handong Global University, and Nathan Kowalsky for proofreading this text and making it more readable.

References

Alexander, Jennifer Karns. 2008. *The mantra of efficiency: From waterwheel to social control.* Baltimore: Johns Hopkins University Press.

Benello, C.George. 1981. Technology and power: Technique as a mode of understanding modernity. In *Jacques Ellul: Interpretive essays*, ed. Clifford G. Christians and Jay M. Van Hook, 91–107. Chicago: Illinois University Press.

Carpenter, Stanley R. 1983. Alternative technology and the norm of efficiency. *Research in Philosophy and Technology* 6: 65–76.

Cózar, José Manuel de. 2000. Toward a philosophical analysis of efficiency. *Research in Philosophy and Technology* 19: 87–100.

Cózar, José Manuel de. 2005. Efficiency. In *Encyclopedia of science, technology, and ethics*, ed. Carl Mitcham, 601–603. New York: Macmillan Reference.
Ellul, Jacques. 1964 [1954]. *The technological society*. Trans. J. Wilkinson. New York: Vintage.
Ellul, Jacques. 1967 [1948]. *The presence of the kingdom*. Trans. Olive Wyon. New York: Seabury Press.
Ellul, Jacques. 1980 [1977]. *The technological system*. Trans. J. Neugroschel. New York: Continuum.
Ellul, Jacques. 1990 [1988]. *The technological bluff*. Trans. Geoffrey W. Bromiley. Michigan: Eerdmans.
Feenberg, Andrew. 2010. Ten paradoxes of technology. *Techné* 14(1): 3–15.
Fogel, R.William. 1994 [1989]. *Without consent or contract: The rise and fall of American slavery*. New York: Norton.
Fogel, R.William, and Stanley L. Engerman. 1995 [1974]. *Time on the cross: The economics of American slavery*. New York: Norton.
Jonas, Hans. 1979. Toward a philosophy of technology. *The Hastings Center Report* 9(1): 33–43.
Mitcham, Carl. 1994. *Thinking through technology: The path between engineering and philosophy*. Chicago: University of Chicago Press.
Pinch, Trevor J., and Wiebe E. Bijker. 1987. The social construction of facts and artifacts: Or how the sociology of science and the sociology of technology might benefit each other. In *The social construction of technological systems: New directions in the sociology and history of technology*, ed. Wiebe E. Bijker et al., 17–50. Cambridge, MA: MIT Press.
Postman, Neil. 1999. *Building a bridge to the 18th century: How the past can improve our future*. New York: Vintage.
Robinson, John Bridger. 1982. Energy backcasting: A proposed method of policy analysis. *Energy Policy* 10(4): 337–344.
Sclove, Richard E. 1995. *Democracy and technology*. New York: Guilford Press.
Smith, Cynthia E. 2007. *Design for the other 90%*. New York: Smithsonian Cooper-Hewitt National Design Museum.
Son, Wha-Chul. 2004. Reading Jacques Ellul's *The technological bluff* in context. *Bulletin of Science, Technology & Society* 24(6): 518–533.
Son, Wha-Chul. 2005a. Autonomous technology. In *Encyclopedia of science, technology, and ethics*, ed. Carl Mitcham, 151–155. New York: Macmillan Reference.
Son, Wha-Chul. 2005b. *Modern technology and democracy*. PhD diss., Katholieke Universiteit Leuven.
Vanderburg, Willem H. 2000. *The labyrinth of technology*. Toronto: UTP.
Vanderburg, Willem H., ed. 2004 [1981]. *Perspectives on our age: Jacques Ellul speaks on his life and work*. Rev. ed.. Toronto: Ananci.
Warren, Rick. 2004. *The purpose driven life: What on earth am I here for?* Grand Rapids: Zondervan.
Winner, Langdon. 1977. *Autonomous technology: Technics-out-of-control as a theme in political thought*. Cambridge, MA: MIT Press.
Winner, Langdon. 1986. *The whale and the reactor: A search for limits in an age of high technology*. Chicago: University of Chicago Press.

Chapter 5
Technological Acceleration and the "Ground Floor of Civilization"

Daniel Cérézuelle

Jacques Ellul used to say that technology, or *Technique,* is intrinsically rational, but irrationalities, and sometimes devastating ones, are created when technology is in contact with realities which belong to a non-technological order. For example, when technique interferes with the natural environment it results in environmental disorganization; and when it interferes with society it results in various kinds of social, political and cultural disorganization.

In this paper I shall focus on one of these aspects of cultural disorganization: the creation of new forms of misery and poverty in affluent industrial societies. My description of this aspect of the socially disorganizing consequences of the technologization of modern life dwells at the confluence of two distinct theses about modern society.

The first thesis is directly borrowed from Jacques Ellul, who wrote in several of his books that technology is a de-symbolizing power. Its rapid development has a destructive impact on traditions and more generally on culture, which is essentially symbolic. The second thesis is borrowed from Ivan Illich, who thought that one of the conditions of modernization and of the rise of industrial society, of the commoditization of resources and of the professionalization of work, is the destruction of popular cultures. According to Illich (1981), the modern age can be understood as an unrelenting war against popular cultures and their framework. This war was waged by the state, the clergy of the various churches, then by the professions and their institutional procedures. As a result, popular cultures and autonomous resources, the *commons* or *domaines vernaculaires*, were devastated. Of course, these two theses are not contradictory but complementary. Ellul insists on the structural impact of technique on culture; Illich insists on its political dimension, in terms of dominance of some groups over other groups. I shall develop an

D. Cérézuelle (✉)
PADES (Programme d'Autoproduction et Développement Social),
1, Rue du Pr Saint Gervais, 93500 Pantin, France
e-mail: daniel.cerezuelle@free.fr

H.M. Jerónimo et al. (eds.), *Jacques Ellul and the Technological Society in the 21st Century*, Philosophy of Engineering and Technology 13,
DOI 10.1007/978-94-007-6658-7_5,

Ellulian – structural, non-political – approach to the Illichian theme of the destruction of vernacular culture and resources as a result of the technologization of life. By so doing I hope to contribute to a better understanding of what Ellul called the unavoidable proletarization in a technological and productivist society.

1 The Social Importance of the Non-monetary Economy

For many years I have studied the issue of poverty in contemporary France where, as in many other developed countries, a great number of households become and remain dependent on public welfare programs and private relief organizations. Most economic and sociological studies of poverty are focused on the social relationships which condition the access of various categories of households to monetary resources. This approach to poverty is necessary but not sufficient, since monetary resources are not the only ones which contribute to the social integration of individuals and households and to their well-being. For most of us who live in developed countries, our way of daily life (*mode de vie quotidienne*) remains dependent on our ability to master informal and non-monetary resources, such as do-it-yourself skills, gifts, and the exchange of services and commodities.

From a quantitative point of view, the domestic and non-monetary economy is still very important. For example, in 1974 it was estimated that the French would devote 41 billion hours to paid work and 48 billion hours to domestic or community unpaid work. The historian Fernand Braudel (1979) used to speak of the "economy of the ground-floor" (*économie du rez-de-chaussée*). He insisted on the essential role of this economy, which exists everywhere under or below the market economy and which is of enormous importance. If this economy was taken into account we would have to increase the GNP from 35 to 75 %, depending on the conventional tools chosen for evaluation.

From a qualitative point of view, this non-monetary or vernacular economy is even more important, since activities that belong to the domestic and communal economy provide the basis for the transmission of behavior patterns, know-how, norms, and values. This is where primary socialization takes place. The domestic and communal economy is organized by complex social rules, which are symbolic and contribute to the stability and cohesion of social life. This "ground-floor of civilization" is of crucial importance for the construction of both the social bond and of autonomous and responsible individuals, but it is neglected by economists and sociologists. Most of them consider that it should decline and that such a decline is the normal consequence of the process of modernization and a good opportunity for creating new professions and more jobs. But from an anthropological point of view, we might fear that an attenuation of the sphere of non-monetary production and exchanges might result in a serious educational regression and in the weakening of the integrative capacity of social life. Nevertheless, this prospect is not taken into account by mainstream economists because, at a deeper level, these social "scientists" believe that this sphere of non-monetary economy is naturally inexhaustible, that it has always been there and therefore will always be there. They take it for granted, but in a technological society what was long taken for granted might turn out to be no longer available.

The founders of political economy (with such notable exceptions as Malthus and Stuart Mill) believed implicitly that progress in both production and human well-being could continue indefinitely because water, air and land would always be available. Their theories of economic progress also implicitly presupposed that there would always be adults willing to have children and to educate them, and that people in this respect would be able to share a common set of values. On the basis of these assumptions, one could well believe that technological and economic progress should be promoted as socially positive, since these fundamental natural and social values would always exist.

2 Remarks on the Symbolic Construction of the Self

Here we can rely on the enlightening analysis of Ernst Cassirer (1923–1929) and – more recently – of Gilbert Hottois (1996: 155), who reminds us that in the philosophical tradition the human being is a symbolic animal, a being who dwells symbolically in space and time. Culture is the ensemble of symbolic relationships that mediate between humans and their world. Symbols (for example, words) are physical realities that develop non-physical, non-causal relationships, and are produced by humans, not by nature. Symbols and their relationships give rise to norms which powerfully influence the mind. Each symbolic world, each culture, produces a network of subjective and intersubjective forces. The power of symbols comes from the fact that they stimulate emotions. Although symbols, through excessive emotional attachment, as in religious bigotry or political fanaticism, can become oppressive and can also be rejected, they establish the space of freedom. Symbols are the condition of freedom: Humans can choose only because they can symbolically consider several possibilities that have different meanings and values. Symbols also mediate between humans and themselves, their impulses, desires, actions, and feelings, as well as between themselves and other humans. Symbolization can therefore be emancipatory in two respects: with respect to oneself and with respect to others (Hottois 1996: 168). Personal autonomy, in order to be achieved, requires symbolic mediations and equipment acquired through the informal educational process. This allows the transmission and interiorization of values, the capacity of relating to authorities representing limits and requirements, and the learning of rules of subordination. Before they can be educated by a formal education (i.e., an educational process consciously organized), children must have been previously "civilized" by an informal educational processes, which provides an implicit foundation for any explicit educational activity.

First, there is the informal learning of very basic human abilities such as the control of one's body, the mastery of time, the ability to communicate, emotional self-discipline, and the acceptance of law. Control of one's body makes it possible to establish relationships to things, and grounds our understanding of how physical effort can alter the world. A mastery of time makes it possible to form projects, to plan, to postpone satisfactions or pleasures. This basic achievement connects with an ability to communicate. I distance myself from my immediate experience and feelings in order to express them in words through the conventions of language. The informal learning of linguistic rules imposes some control over emotions insofar as it gives rise to a capacity to place a kind of wall around our aggressiveness, rivalries, desires, and impulses. Without the means to suspend immediate drives, dialogue and relationships are not possible. Finally, the acceptance of law is a central aspect of the construction of a social self. The emergence of the citizen is possible only through an ability to refer immediate drives to a symbolic referee for ultimate judgement. This same ability, in turn, contributes to the framing of our experiences in the structures of a common language, organizing them in time, building projects, learning social codes (of whatever type), and engaging in organized social exchange.

Second, there is the informal learning of how to participate in symbolic patterns of behavior and collective action, the means by which we acquire many of the above-mentioned abilities. These are at the same time quite complex and interrelated, while operating at diverse levels of interactivity. Examples of symbolic patterns include the following:

- The rules of language and the combination of signs.
- The codes of postures, including ritual dances, songs, lullabies, games, etc., which facilitate the sharing and the transmission of experiences (e.g., how to cope with separation, loss, or desire).
- Codes of etiquette and "savoir-vivre" which allow a pacific (albeit often unequal) frame for interaction (Picard 2007).
- Modes of relationship with non-human life (animal and vegetable), which contribute to the understanding of oneself and others (e.g. physical pain).
- Participation in collective and technological systems of action (building, cultivating, domestic life), which allow access to a collective time and to social and ecological responsibility.
- Rules for giving gifts of symbolic exchange.
- Rules governing sexual exchange and incest.
- Models of coexistence with different circles of reciprocity: that is, the nuclear family, extended family, kinship, clan, village, peer group, trade, church, etc.
- Myths, tales and stories that give meaning to experiences and prepare for them.

Finally, third, there is the informal learning of certain character traits. This learning takes place in *multidimensional* ways – as, for example, in the uniting of speech and action, or the interaction of the social, technological, and religious. The emotional dimension gives rise to the *internalization of values*, that is, of limits, duties, authorities. It is at this emotive level that we initially become a self, with an organized behavior, capable of social exchange and of responsibility. Such traits operate in an *unconscious* way, and are thus difficult to describe and analyze. They are not interchangeable, despite the fact that they might be very diverse in time and space. They likewise operate at a slow pace, at the obscure levels of psychic life, by a nocturnal unfolding, thus with very different temporal requirements than found in, for example, rational decision making. They are transmitted as *traditions* from one generation to the next (e.g., how to handle a baby). The reproduction of this symbolic capital depends very much on the vitality of a non-monetary economy and of community life (hence the African saying "it takes a whole village to raise a child"), both of which nourish the symbolic dimensions of doing and exchanging.

This "civilizing process" (Elias 1978) makes possible the emergence of "selves" structured on the basis of incorporated values that allow them to respond meaningfully and effectively to their experiences and the circumstances of daily life.

3 Symbolic Resources Are Not Easily Renewable

The study of contemporary culture suggests that the transmission of the skills and know-how of daily life, which play an important role in the construction of personality and of social capabilities (in the sense of Amartya Sen [2000]), does not function properly. Two hundred years after the beginning of industrial, technological civilization, we discover that some of the natural resources we depend on are non-renewable. But this is only one side of the story. The other is that important cultural resources, too, are not easily renewable. The subtle processes that organize human life – at both collective and individual levels – are fragile, and for many people the informal processes of transmission and incorporation of social and practical knowledge are less and less effective. I shall briefly mention three instances of the weakening of these processes.

3.1 The Erosion of Civility

In a recent poll in France, people were asked what was the main cause of anxiety in their personal life. Far ahead of the usual concerns about money, exhausting commuting situations, noise or lack of time, the main concern of a surprising large number of people (60 %) turned out to be the lack of politeness in the circumstances of daily life (Kremer 2012). More and more educators say that they are now dealing with young people who have not mastered the conventions of peaceful social intercourse. Because of the lack of training in due time, students do not master the psychological and symbolic resources necessary for controlling their emotions, and for organizing their conduct in an educational setting. Incivility and violence become a pervasive problem, and many educational institutions, in spite of increasingly sophisticated or "adapted" strategies, can no longer function properly (Ernst 2011).

3.2 The Loss of Dwelling Skills

Many people today have had no introduction to the know-how and behaviors that allow us to *dwell* somewhere. Others who once possessed such capabilities have lost them. As a result many lodgings are not properly maintained and some housing units are so neglected and dilapidated that they must be destroyed.

3.3 The Rise in Nutritional Disorders

In many affluent industrial societies an increasing number of adults do not know how to feed themselves and how to feed their children. As a result, obesity and diabetes are becoming major public health problems, spoiling the life of countless individuals. The impact on both the public and private health budgets will be huge.

3.4 Civilizational Context of This Weakening

Three basic features of modern life in a technological society contribute to the erosion of our symbolic endowment.

First, accelerated change in all aspects of life has two consequences: on the one hand, a disqualification of traditions, which were elaborated in a very different context; on the other, a great difficulty in creating new traditions. As soon as such traditions begin to emerge they are immediately outdated (the collapse of working-class culture in Europe is a good example). As a result, each generation starts anew, at a great cost (the problem of eating habits among the lower classes in the US is one sad example).

Second, monetarization and commodification of economic life erodes the importance of non-monetary relationships. It is possible, indeed, to view this evolution as an emancipatory one: once I get paid for my work, then I owe nothing to anyone, and no one owes me anything; therefore I am free to spend my money and live as I choose. But this emancipatory dimension of the monetary economy should not blind us to the fact that it is at the level of non-monetary economy that many important socializing and symbolic processes take place, and that a complete de-vitalization of this non-monetary economic sphere may hinder the reproduction of our social symbolic capital. As Marx noticed long ago, money is a powerful agent of de-symbolization. In the world of modern capitalism, monetization of human relationship has no fixed limits. The dynamism of capital constantly displaces the symbolic norms that regulate the relationships between humans or between humans and the world and deprives them of their organizing power, disqualifies, and finally abolishes them. In modern society, everything has a price: sorrow, parts of the living body, the environment etc. can all be reduced to the status of commodities.

Third, technoscience is also a powerful agent of de-symbolization. It seems that the unfolding of operative power contributes to the weakening of symbolic points of reference. Like money, technology is a powerful social operator that tends to free itself from symbolic constraints. The realm of technological operativity constantly expands and therefore displaces ethical norms, a process exemplified by the development of biotechnologies and techniques of human reproduction. There is no intrinsic teleology in the world of science. Nothing is untouchable; everything can be acted upon and modified by a calculus of technical operations.

4 On Some Anthropological Limits of the Technicist and Economicist Ideal

Thus the rapid change of the technological and economic infrastructure of social life disqualifies and renders obsolete and inefficient many of the symbolic resources that organize individual and social life. Besides, the creation of new symbolic equipment, of new traditions, is not easy. By necessity, the rate of creation of such resources is slow. The time unit is at least one or several generations, and it is doubtful that it might be accelerated, as exemplified by the failure of modern revolutions to create a "new man" (we shall return to this later). By contrast, the tempo of technological and economic change is constantly accelerating. This discrepancy is a powerful cause of the social and cultural disorganization, which, in the nineteenth century, Marx had already mentioned with respect to the development of the monetary economy. But Marx did not draw all the consequences from his observations. Being a philosopher, Marx was mostly interested in the formal and intellectually elaborated dimensions of the symbolic resources of social life: law, art, religion, and philosophy. As the heir of a rationalist tradition, he neglects the informal dimensions of the symbolic resources and their importance in the construction of social bonds, and he also underestimates the specificity of non-monetary relationships and their importance in economic life.

In a context of rapid social change, not only some of those who have access to those informal resources experience difficulties at using them in their daily lives, but we can observe a common failure of their transmission from one generation to the next. This failure (*panne*) concerns both the skills necessary for living together (sociability, civility) and the practical know-how of daily life. Some will object to the argument for the de-symbolizing role of technology that the symbolic foundations of a society can renew themselves. Are we not living in a world which, thanks to technology, exhibits a huge symbolic production? Although this is true, we should also take into account two obstacles that hinder the organizing power of the peculiar form of symbolic production associated with modern technological and economic development. One results from the tempo of the production, the other, its contents.

In order to be socially effective, the creation, transmission and internalization of symbolic forms needs to take place with a special, slow tempo. Taking time is a necessity. Yet in the contemporary world individuals experience an incessant flow of new symbols and signs, the meanings of which are quickly outdated and fade away. What characterizes an inflationary situation is, precisely, the rapid loss of value and the efficacy of signs and messages, exorcized by their very proliferation. This rapid tempo of production and obsolescence is a symptom of a loss of meaning and efficiency, which we can observe in the realm of academic philosophy (Hottois 1979) as well as in the culture of daily life. The contemporary inflation of signs and images is incapable of creating the symbolic world described by Cassirer. On the contrary, it contributes to its organizational anemia and loss of emotional pregnancy. This failure in the transmission of the practical and social skills of daily life will remain a persistent feature of technological society.

Several authors further suggest a second and more fundamental limit of contemporary symbolic production. They do not think that the world today is characterized by a de-vitalization of the symbolic order. Quite the opposite, they acknowledge that technological change contributes to the de-vitalization of the symbolic capital inherited from earlier stages of development, but claim that technology is now endowed with a powerful symbolic value of its own. Several symptoms suggest that the symbolic world specific to technological society is a cause of disorganization, not negatively because of a default of productivity and durability, but positively, because of its own orientations and implicit values. This symbolic world differs from historically earlier ones because it has a self-destructive dimension. By rejecting the very principle of limit, the spirit of technicism hinders the process of construction of the autonomous self, which technological society actually needs. The unfolding of the symbolic world that corresponds to the process of technologization may well weaken the anthropological basis that makes technologization possible. This is the contention of authors such as Jean Brun (1970), Luc Boltanski and Eve Schapiello (1999), Alain Supiot (2005), and Jacques Généreux (2006). For Cornélius Castoriadis, a socialist philosopher who was a friend of Ellul, the politico-economic system that is replacing traditional capitalism can only reproduce itself thanks to anthropological types inherited from previous ages. The system can reproduce itself only insofar as it can rely on individuals whose values are not the values of the system. This is the paradox: modern science, just like modern economy, destroys the world of symbolic values necessary for the social production of the educated, civilized, and somehow moral agents it cannot dispense with. Capitalism needs entrepreneurs who respect contracts; it needs judges who have more reverence for law than for profit. Science and technology require scientists and engineers who do not fake data and have some sense of respect for the safety, at least of their colleagues.

5 From the Technological to the Symbolic Order

The symbolic dimensions of man's relationships to his techniques are crucial for understanding some of our contemporary problems. Contrary to intellectualist and utilitarian interpretations, in spite of the rationality of the knowledge and of the means which make it efficient, the technical action of *homo faber* is not simply organized by objective ideas and concepts. This action is always rooted in a symbolic "ground." Undoubtedly, technical action aims at mastering a matter by means of operations, which lend themselves to an objective description, but neither the matter nor the aim of this action are completely objective. Cassirer reminds us that it is the symbolic form which makes possible the constitution of a significant matter. Outside of this form, the matter has no meaning. Meaning is first; the symbolic form gives a meaning to the elements of reality; it organizes action and knowledge. This is true not only for scientific knowledge but also for technical action and the culture of daily life. The utilitarian approach to technology, which is misleading insofar as it assesses technology in terms of objective needs, naturally elicits a technical response associated ways of thinking. According to Cassirer, the symbolic form does not stem from necessity; it does not respond to a pre-existing need. In order to understand technological evolution, we should renounce an evolutionary model of cumulative, incremental, linear progress and keep in mind that our techniques are not the mere application of objective and existentially neutral knowledge and procedures. Cassirer's philosophy helps us to understand how techniques are not created and transmitted through a logical and utilitarian pattern, but rather they are embedded in a web of symbolic relationships that constitute a coherent ensemble, a symbolic world, from which it is difficult to extract them, since this ensemble must be transmitted as a whole. Each symbolic form allows a specific access to the world according to a perspective that is never purely objective and elicits the commitment of the self in the world in a way that is sensual, emotional, and carnal as well as intellectual.

Integration in a symbolic form characterizes techniques in pre-modern societies; it also characterizes techniques of daily life – of the ground floor of civilization – in modern ones. According to the French sociologist Jean-Claude Kaufmann (1997), whenever we clean our home, wash and iron our clothes, or prepare our meals, the body and our emotions play an essential role in the workings of "domestic reason." Domestic action depends on rituals; cleaning the home implies the performance of something like a dance, whose rhythm organizes our motion and gestures. Technical action cannot be separated from a habit and from a life-history which results in a constant interplay between mind and body. Together with objective and coldly calculating intelligence, a sensitive, emotional, and intuitive intelligence is at work. In order for that intelligence to be used, tools must be incorporated. Habits and the memory of these rhythms which organize technical action and the use of tools are stored in the body. Hence these "initiating dances" are necessary for getting ready – mind and body – for action. Then, most of our gestures are enacted in a state of semi-consciousness or "mental slumber" from which the self rarely awakens, except when facing an unpredicted situation. Feelings play a decisive role, especially when we do not act "properly," which is experienced first at the emotional level, or when inaction is experienced as shameful. In all these situations, a symbolic background defines the meaning of technical acts and facilitates the incorporation of tools.

The pursuit of development may cause an important erosion of the sphere of activities and relationships, which is one of the informal bases of our symbolic capital. In a context of rapid technological change many individuals who had been equipped at the right time of their life with appropriate practical know-how and symbolic resources cannot put them into practice, because their environment has changed. As a result the transmission from one generation to the next is becoming less effective. The emergence and resilience of new forms of misery can be understood as a warning about the limits of the technologization of life and of the universalization of monetary relationships: they favor a rapid social change which tends to destroy the very anthropological basis which makes this change possible. But renouncing the technicist and consumerist ideal and imposing a slower pace on technological innovation is extremely difficult for *homo technologicus*. We are now immersed in a technicist and productivist mythology or "imaginaire," which we might call, in a Weberian way, *the spirit of technicism,* which hinders our capacity to address these problems. Replacing, by means of reasoning, false ideas by more adequate ideas does not work, since too often, as Jean-Pierre Dupuy (2002) warns us, we do not believe what we know when what we know contradicts this spirit of technicism. Ernst Cassirer tells us that the symbolic form is both an element of our environment and the means by which we can assess this environment and determine what is real, true and important. Facing the problems created by technological acceleration, *homo technologicus* spontaneously responds by more technique. Demythologizing this spirit of technicism is therefore of the utmost political importance.

References

Boltanski, Luc, and Eve Schapiello. 1999. *Le nouvel esprit du capitalisme*. Paris: Gallimard.
Braudel, Fernand. 1979. *Civilisation matérielle, économie et capitalisme, XVe-XVIIIe siècles*, 3 vols. Paris: Armand Colin.
Brun, Jean. 1970. *Le retour de Dionysos*. Paris: Desclée.
Cassirer, Ernst. 1923–1929. *The philosophy of symbolic forms*. New Haven: Yale University Press.
Dupuy, Jean-Pierre. 2002. *Pour un catastrophisme éclairé*. Paris: Le Seuil.
Elias, Norbert. 1978. *The civilizing process*. New York: Urizen.
Ernst, Cécile. 2011. *Bonjour madame, merci monsieur: L'urgence du savoir vivre ensemble*. Paris: JC Lattès.
Généreux, Jacques. 2006. *La dissociété*. Paris: Le Seuil.
Hottois, Gilbert. 1979. *L'inflation du langage dans la philosophie contemporaine*. Bruxelles: Editions de l'U.L.B.
Hottois, Gilbert. 1996. *Entre symbole et technoscience, un itinéraire philosophique*. Seyssel: Champ Vallon.
Illich, Ivan. 1981. *Shadow work*. London: Marion Boyars.
Kaufmann, Jean-Claude. 1997. *Le cœur à l'ouvrage, théorie de l'action ménagère*. Paris: Nathan.
Kremer, Pascale. 2012. Soyons polis! *Le Monde,* 4 Janvier 2012.
Picard, Dominique. 2007. *Pourquoi la politesse? Le savoir-vivre contre l'incivilité*. Paris: Le Seuil.
Sen, Amartya. 2000. *Development as freedom*. New York: Anchor Books.
Supiot, Alain. 2005. *Homo juridicus, essai sur la fonction anthropologique du droit*. Paris: Le Seuil.

Chapter 6
Technological System and the Problem of Desymbolization

Yuk Hui

Who is Jacques Ellul? Prophet, sociologist, philosopher, theologian? How should we read an author who has taken on such a multiplicity of roles in a career of prolific productivity? Shall we read using the theoretical frames he set himself, challenge him with postmodern theory, or link his theory to the different schools that characterize contemporary discussion? My reading takes the third approach, adapting the work of Gilles Deleuze, to create consistent concepts that allow us to renew our understanding of reality. It is undeniable that technology itself is the source of a transformation of reality, and this prompts us to constantly rethink the milieu in which we are living. This essay focuses on what Ellul calls a technological system, especially on the aspects of symbolization and desymbolization that characterize the technological evolution which separated human from nature. On the one hand, this system characterizes a permanent departure. It takes human beings to the middle of the sea, where they can no longer identify their own land, nor can they reach the horizon which had seemed to be so close, to paraphrase Nietzsche from *The Gay Science*. On the other hand, the separation presents us with contemporary situations that bear their own specificities and pose risks that must be tackled individually and in detail.

Ellul's conceptualization of a technological system suggests a new way to mediate the relation between human beings and, following the vocabulary of Gilbert Simondon, technical reality. The technical reality constitutes the world in which we dwell, an existential analytic (if Heidegger's project still holds its importance today) that can only be reinvented by admitting that we are actually beings-in-the-technological-system. But it is also essential to evaluate the technological system according to a

Y. Hui (✉)
Centre for Digital Cultures, Leuphana Universität Lüneburg,
Scharnhorststr. 1, Lüneburg 21335
e-mail: yuk.hui@leuphana.de

Institut de Recherche et d'Innovation du Centre Pompidou,
Rue Aubry le Boucher, 4, Paris 75004, France
e-mail: yuk.hui@iri.centrepompidou.fr

H.M. Jerónimo et al. (eds.), *Jacques Ellul and the Technological Society in the 21st Century*, Philosophy of Engineering and Technology 13,
DOI 10.1007/978-94-007-6658-7_6, © Springer Science+Business Media Dordrecht 2013

technical reality that never remains static, and this requires reinvestigating Ellul's technological system in contemporary terms.

This essay is divided into three parts. The first discusses the relation between a technological system and desymbolization in Ellul's thought. The second evaluates the technical system and desymbolization through a discussion of Ellul's inspiration by Simondon, looking at the technological system we confront today. The third examines the technical system from the perspective of another French philosopher, Bernard Stiegler, who was also inspired by Simondon. These steps help bring Ellul's thought to bear on contemporary technical reality.

1 Evolution of the Technological System

My reading of Ellul will thus focus on his *Technological System,* which was published in 1977 and translated into English in 1980. In this book, Ellul proposes technology as both an environment and a system. An environment is easier to understand, since it is indicated by the artifacts that surround us everyday. A technical system presents something different. Considering the constantly evolving technical system, Ellul proposes that it is useless to talk about a single technology, but rather that we must grasp the technological system as a totality. A technical system for Ellul is made up of the technical phenomenon and its progressions. A progression is not what people commonly understand as evolution through time, but rather a vital force within the objects themselves that constitutes their progress from one stage to another. The technological system in this sense is no longer a collection of objects or technologies, but rather a gigantic force that pushes forward the technical lineage. One of the key consequences that Ellul identifies with such technological progress is the process of desymbolization. Put simplistically, the evolution of a technological system is characterized by a dialectical movement between the destruction of old symbols and the creation of new ones. This may sound similar to Ernest Cassirer's well-known proposal that culture is a constant movement between *forma formata* (structured structure) and *forma formans* (structuring structure) (Vandenberghe 2001), but Ellul's theory is distinct from Cassirer's. The relation between desymbolization and the technological system is one of the more interesting but least developed points in Ellul's theory.

Consider now the meaning of symbolization and desymbolization in this context. Commenting on the relation between the technological system and rituals, Ellul proposes that

> the function of symbolization no longer attests to a specifically human power. It is now subordinated to a different order, a different function, which are both already created by man. And if that function is performed, it proves that technology is now the true environment of man (otherwise, he would not feel the need to operate with symbols in this connection) (Ellul 1980 [1977]: 177).

We can easily recognize this concept of desymbolization in an anthropological sense. Symbolization is a process that creates association between human and nature, gods, or spirits through artificial objects such as totems, figurines, and more.

As Ellul illustrated, in certain civilizations it was forbidden to work on the ground with iron tools since nature was conceived as mother and iron tools were considered harmful to the mother.[1] The symbol of earth as a mother figure is transcended when a technological system is adapted due to different cultural factors, such as war and famine. Symbols that were once mediated between different powers and were included in ritual practices are eliminated in the process of technological development. Desymbolization is thus a process of short-circuiting that brings forth an efficient and automatic technological system in exchange for the traditional values and forms of life.

Nevertheless, this is too simplistic a reading of Ellul. Instead we should go back to Simondon, who directly inspired Ellul's concept of the technological system. By harkening back to Simondon, we can discover some latent aspects of desymbolization in Ellul's thought. This proposal is also in response to Ellul's proposition that in order to study the technological system, one must go inside the technological system and its specificity. Ellul's debt to Simondon is obvious in *The Technological System*, where he quotes Simondon extensively, especially in the chapter on "Technology as a System," where he repeatedly references *On the Mode of Existence of Technical Objects*, Simondon's doctoral dissertation from 1958.

But to begin, note something concerning the English translation of *The Technological System*. Ellul term is *le système technicien*, which literally means "the technician system." What I understand by Ellul's use of *technician* is this: that we are living in a culture that depends on technical reason, which is no longer constrained by moral or religious judgment. Technicians are producing a culture with technical reasons. Hence culture is more technical than technological, if by technological we refer to infrastructures, machines, and all kinds of artificial objects. The translation of technological system should not be understood merely as an ensemble of artificial objects, but includes reason operating within technical constraints.

On the Mode of Existence of Technical Objects proposed what Simondon calls a "mechanology." "Mechanology" investigates the existence of technical objects through their lineage toward perfection. Mechanology suggests that the traditional conceptualization of technologies as in opposition to culture is mistaken; instead, culture is technological and technical. Simondon describes this lineage from the origin of technology to the point where it provides an increasingly concrete object through the example of the Lee de Forest triode. The triode is an evolved version of the diode, a device that controls the flow of direct current. In the simplest diode vacuum tube the cathode is heated and hence activated to release electrons; the anode is positively charged so that it attracts electrons from the cathode. If the voltage polarity is reversed, the anode is not heated and thus cannot emit electrons so that no current passes through. A triode places a grid between the anode and the cathode; a direct current (DC) can give a bias to the grid: if negative, it repels some of the electrons back to the cathode and hence serves as an amplifier. Simondon proposes that the origin of the triode is not the diode but "the condition of irreversibility of the

[1] Jacques Ellul, 1992. *La trahison de la technologie*, video. http://www.dailymotion.com/video/xczyxj_jacques-ellul-le-systeme-technicien_webcam

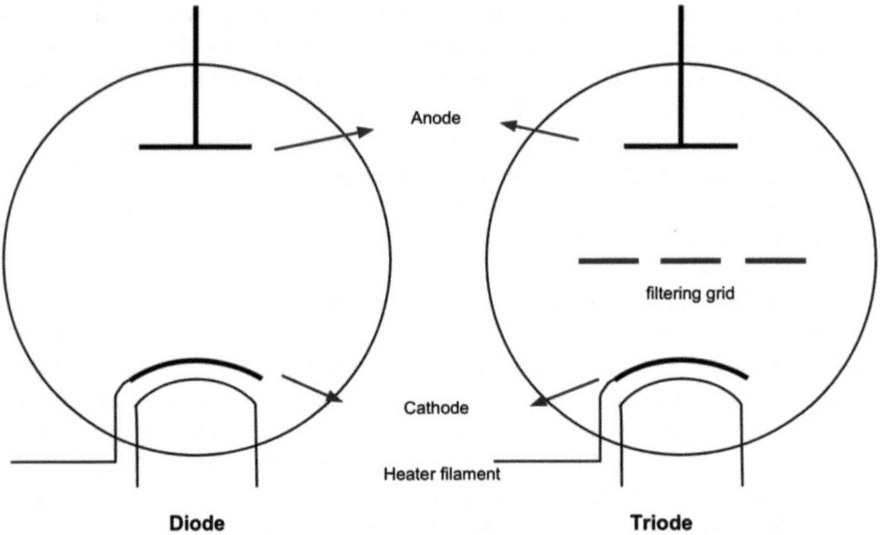

Fig. 6.1 An indirect heated vacuum tube diode and triode (Illustrated by the author)

electrodes and the phenomenon of the transport of electric charges across the vacuum" (Simondon 1980: 36) (Fig. 6.1).

A technical individual is a technical object that incorporates or adapts an external milieu into its functioning. This external milieu is what Simondon calls the "associated milieu" or environment that becomes part of its functionality. For example, Simondon often referenced the Guimbal turbine (named after the engineer who invented it). To solve the problem of overheating and energy loss, it uses oil to lubricate the engine as well as to protect its parts from river water, which it uses as a cooling agent (Simondon 2005). The river here is an associated milieu insofar as it is part of the system but not a component in the machine. Simondon's approach to technical objects differs from that of previous philosophers and phenomenologists in that he does not reduce the technical object to the intentional product of consciousness but makes it an object to be examined in its own right. He proposed to study the genesis of the technical object itself, less in a biological sense than in a mechanical one. A technical object regains its materiality and attains a different degree of concreteness or perfection in contrast to what cybernetics terms "control." Technical objects form ensembles; they also create a secondary associated milieu, which maintains the connectivity and metastability of the technical ensemble. Technical ensembles or groups of technical ensembles constitute what Ellul calls a technological sub-system. An example of this would be transportation technology, including the road infrastructure, signs, and more. Such sub-systems then further form the basis of a technological system.

The significance of seeing the technological system in this way is that we can further discover the desystemization process as the materialization of different connections between different technical ensembles. The process of desymbolization involves the creation of a new kind of materiality. Connections are realized, for example, between the pulleys and the wheels in a mechanical system, and between the optical cables or electronic wires in a modern electronic apparatus. Desymbolization must be seen as the emergence of materiality that compensates for the weakness of the traditional form of mediation, and promotes the concept of control and efficiency. The technical system is in constant struggle for a common ground that allows it to establish material connections. The cybernetic movement in the last century attempted to find common ground in logic, information, and signals that could integrate human beings into a technical system. Finally the system will subject all elements to control. Ellul is not unaware of this, as when he praises Simondon:

> Simondon excellently demonstrates this process of *causal evolution* on multiple levels. First of all, as the technological object evolves, it suppresses secondary effects which may prove to be obstacles and specializes each structure as a 'positive synthetic functional unit': 'The concrete technological object is one that is no longer struggling with itself, one in which no secondary effect damages the functioning of the whole.' Thus, technology itself evolves by eliminating, in its own movement, anything that hinders it from being perfectly realized; this is a progression with no external objective (Ellul 1980 [1977]: 275).[2]

The process of "elimination" in its own movement is what we just mentioned above. The production of a new materiality bypasses the domination of the old one, just as manual labor is replaced by electrically-driven mechanical forces, symbolic mediations is replaced by direct control. Hence Ellul concludes:

> The results are: escape symbolization, as in modern art; artificial symbolization (bearing upon technology but perfectly useless and meaningless, as we shall see later on). The approach to, the grasp, interpretation, and control of, the technological environment cannot take place through symbolization. As for the natural environment, symbolization is made perfectly meaningless here by the dominance of utilitarian technology (Ellul 1980 [1977]: 40).

2 Data Processing and Technological System

Before we go to the third aspect of desymbolization, we must renew our understanding of the technical reality. We have to pose the question: what characterizes the technological system today? Or more precisely: what is the new materiality that produces a unified technological system? We can answer that it is the production and processing of data. In fact, by the end of the 1970s Ellul already identified the significance of data processing as a force that carries out further extensive desymbolization, far before the advent of the Internet. He said:

> Thanks to the computer, there emerged a sort of internal systematics of the technological ensemble, expressing itself by, and operating on, the level of information. It is

[2] Simondon's "objet technique" is often translated as technical object, and sometimes adopted and translated as technological object, as in this quote reproduced from Ellul, but we have to bear in mind that when Ellul talks about technological objects by referring to Simondon, it is what we call "technical object" in this article.

> through reciprocal total and integrated information that the subsystems are coordinated. This is something that no man, no human group, no constitution was able to do (Ellul 1980 [1977]: 102).

It is even more compelling when we noticed that in the late 1970s, Ellul also talked about closed data and open data, a topic that was put on the agenda by the European Union 30 years later.[3] Although Ellul talked directly about information in *The Technological Bluff* (published in 1990), but it was only in his discussion of data processing that I think Ellul grasped the materiality of the contemporary technological system. Of course when Ellul was working, data processing was limited to a small number of computers and only a few data analysis experts. Today, data has become the key question for technological development in different industries, especially with the advance of the Internet. With the proliferation of personal computers and Internet access, data production has become ubiquitous and is no longer limited to experts. Here let me quote the UC Berkeley computer science Professor Michael Franklin about the production of data by a single user, from which we can peep into the universe of data with which we live:

> Most tweets, for example, are created manually by people at keyboards or touchscreens, 140 characters at a time. Multiply that by the millions of active users and the result is indeed an impressive amount of information. The data driving the data analytics tsunami, on the other hand, is automatically generated. Every page view, ad impression, ad click, video view, etc. done by every user on the web generates thousands of bytes of log information. Add in the data automatically generated by the underlying infrastructure (CDNs, servers, gateways, etc.) and you can quickly find yourself dealing with petabytes of data (quoted by Lorica 2009).

On the other hand, we must be aware that the production of data is not limited to user-generated content, for example those the users consciously contribute to search engines and social networking websites such as Google, Facebook, etc. In fact, data collection has also become ubiquitous. Different institutions devoted to the natural sciences and the medical sciences, for example, are producing large amount of online data ranging from the records of patients to protein structure, allowing them to better understand different patterns and to produce simulations. There are also emerging sets of big data which are not consciously produced by users but are collected using different sensors, such as GPS and RFID, etc. This type of data can be perceived as the "unconsciousness" that discloses hidden patterns of human/animal behaviors. All these means contribute to an emerging digital milieu and a concretizing technological system, in which different entities can be digitized and thus connected by data links.

In recent years we heard a lot about the "Internet of things." These data are not raw data in the sense that they are formless; instead, these data are formalized by

---

[3] See the European Commisioner Neelie Kroes's discussion on open data. http://blogs.ec.europa.eu/neelie-kroes/opendata/2001. Accessed 8 June 2012.

Album• Application• Checkin• Comment• Domain• Event• FriendList• Group• Insights• Link• Message• Note• Page• Photo• Post• Review• Status message• Subscription• Thread• User• Video

Fig. 6.2 List of objects in the Facebook graph API (Facebook developers) https://developers.facebook.com/docs/reference/api/. Accessed 17 May 2012

different metadata schemes. Metadata, literally data about data, are the sources that establish these connections. For example, consider the book *Technological System.* Its metadata consists of title, author, page number, ISBN number, publisher, years of publishing, etc. The more detailed the metadata schemes are, the more connections are established. It is also fair to say that data are taking a more concrete form than Ellul imagined in the 1970s. These digital entities with formalized metadata are what I call digital objects, in a conceptual renewal of Simondon's idea of technical objects.

If we take computation to be a cognitive process as defined in the works of Alan Turing, John von Neumann, and Warren McCullough, etc., algorithms and databases are mechanisms that govern cognitive processes, and data are literally treated as "objects" by computers. Hence the founder of the World Wide Web, Tim Berners-Lee, who proposes the formalization of metadata in the name of the semantic web, is able to call such a technological system a global mind (Berners-Lee 2000). Human beings are also reduced to computational processes, and ultimately digital objects. Digital objects thus become the basic units recognized by both computers and human users. This is not simply a philosophical conceptualization. If we look at the Graph API that defines the core data structure of Facebook,[4] we are not surprised to find out that all the elements are defined by the Facebook engineers as objects (Fig. 6.2).

Facebook is composed of these formally defined objects. The idea behind the Facebook Graph API is to establish connections between different objects. For example we can see intuitively that every album has photos, and every photo has comments. A comment consists of attributes like author, timestamp, and message among other things. Another core concept is the Open Graph Protocol that allows users to create connections between different platforms. By clicking "Like" in another website, Facebook and its partner website will have the data and be able to produce a graphical analysis of a user's social metadata. The aim is to create data-networks which allow these social networking websites to create relevant contexts for the users. In other words, networks are composed of digital objects, which are in turn defined by multiple layers of metadata. Their appearances depend on complicated systems of relations and algorithms that are not accessible to the users who interact with them. These are new types of industrial objects not yet properly addressed in the work of theorists of technological society.

[4] See https://developers.facebook.com/docs/reference/api/. Accessed 17 May 2012.

3 Technological System as Retention Systems

We have discussed before that desymbolization does not only eliminate symbols, but also produces new symbolizations. Commenting on the process of desymbolization by the imposition of technological system, Ellul describes the new symbols in the contemporary capitalistic society:

> On the one hand, man's inherent power of symbolizing is excluded; on the other hand, all consumption is symbolic. The technological system is a real universe, which constitutes itself as a symbolic system. With respect to nature, the symbolic universe was an imaginary universe, a superordinated reflection, entirely instituted by man in relation to this natural world (Ellul 1980 [1977]: 177).

Consumption is nevertheless a very limited phase of the dialectical process of symbolization and desymbolization. By describing consumerism as the totality of the new symbolization Ellul seems to ignore the question of materialization discussed above. This is what makes "desymbolization" a "problem," as identified in the title of this article. Consumption as symbolization is to a large extent psychological and psychical, more and more motivated by moving images, sounds, and different technological apparatuses. If one is using Facebook, the advertisements that pop up to the users are already determined by the data that represent the browsing history of the users. That is to say, behind consumption is another dimension that has been overcome by the concretized data network. What then are the new implications of our current technological system? It will also be too easy to call it a total control or to follow Gilles Deleuze, who calls it the "control society" that gives way to cybernetics. The remaining task is to understand the mechanism behind this control, to look into the technological system in its details. Here I propose to create a link between Ellul's prophecy and the work of another French philosopher, Bernard Stiegler, who was largely inspired by Simondon.

Bernard Stiegler (2010) calls both technical objects and technical systems *tertiary retention*. Here we may differentiate between two types of tertiary retentions: one the "already there" (following Heidegger) of the world that is already a technological system, as exemplified by the history and material conditions in which we already live. The other is the exteriorization of memories, which was realized in writing, printing, analogue technologies and now digitization. Data processing is one of the most important results of digitization.

The word tertiary retention is a supplement qua a critique to Husserl's understanding of time-consciousness. To explain Husserlian time-consciousness, let us imagine that we are listening to a melody. We are experiencing a flux of consciousness, which is the passing of the now. The now that is retained immediately in my mind is what Husserl calls primary retention, the melody that I can recall tomorrow is called secondary retention; these retentions also condition the protentions, which include anticipations and projections of the future. Tertiary retention supplements the finitude of the first two kinds of retention with an infinite repertoire of memories, made possible by digitization. But tertiary retention is also the source of primary retention, and the support of the secondary retention is also the source of protention.

In this sense, we can see a third sense of desymbolization on top of the first two kinds of desymbolization discussed above, regarding demystification and the materialization of relations within the technical system. Since now the mediation process is subject to the control of retentions, what happens to the symbols when they are no longer a simple transformation from one form to another, but take a more radical move? Hence Bernard Stiegler and Irit Rogoff write, digital technology

> creates a new organization of the circulation of the symbolic. Within this new mode of organization, suddenly the production of the symbolic becomes industrial, subject to industrial processes. Here you encounter the production of symbols on the one hand, and the consuming of such symbols on the other – an aporia because it is impossible to consume a symbol. The symbol is not an object of consumption; it is an object of exchange, of circulation, or of the creation of circuits of trans-individuation. So this situation suddenly produced what I call short-circuiting – of transindividuation (Stiegler and Rogoff 2010).

Fully appreciating this quote would require examining both what Simondon calls individuation and the concept of transindividuation further developed by Stiegler, however this would take the article in a different direction. What we can take from this is that the process of desymbolization and resymbolization, which is also materialization and imagination, no longer operates on the level of signification in linguistic terms. What used to be a signification process within the mind now can be short-circuited by the manipulation of the tertiary retentions, which are digital objects or data. Desymbolization brings humans and machines into a symbiosis, a new nature that is largely overlooked in the classical opposition between nature and technics. What happens in this aspect of desymbolization is not the loss of meanings or references, but the alteration of meanings produced by the new circuits. Symbols cease to be merely representations, but come instead to contribute to the controlling functions of the technological system, in which human and machines are interconnected circuits.

4 Conclusion

The above exposition attempts to bring out the three aspects of desymbolization brought about by the evolution of the contemporary technological system. First, there is deritualization in an anthropological sense; second, the materialization of relations; and third, the creation of circuits within the retentional system that is also part of the technological system. These first two points are briefly mentioned in Ellul's *Technological System* but are not fully developed. The third point to integrate Ellul's commentary on data processing with the contemporary situation of desymbolization. The merit of Ellul's theory is not simply his prophecy but more importantly his attempt to outline the technological cycles that transform our culture and the ontogenesis of human beings.

Desymbolization is a general effect of technological development, as we saw at the beginning of this article regarding Cassirer's proposition on symbolic forms. It is also a process of the concretization of technical objects, the materialization of technical reasons, and the adaptation of milieus into an expanding technological system. It is no coincidence that for Ellul, Simondon, and Stiegler, the question of capitalism today is not about capital in an economic sense, but rather about machines (Chabot 2003; Jézéquel 2010; Stiegler 2010). Or more precisely, the technological system. The understanding of technological systems and their inner dynamics is crucial to analyzing and problematizing understandings of contemporary culture. Ellul's *Technological System* remains an important place to start.

References

Berners-Lee, T. 2000. *Weaving the web: The past, present and future of the world wide web by its inventor*. London: Texere.

Chabot, Pascal. 2003. *La philosophie de Simondon*. Paris: Librairie Philosophique J. Vrin.

Ellul, Jacques. 1980 [1977]. *The technological system*. London: Continuum.

Ellul, Jacques. 1990. *The technological bluff*. Grand Rapids: Eerdmans.

Ellul, Jacques. 1992. *La trahison de la technologie*, video http://www.dailymotion.com/video/xczyxj_jacques-ellul-le-systeme-technicien_webcam.

Jézéquel, Jean-Pierre. 2010. Jacques Ellul ou l'impasse de la technique. In *Revue du MAUSS permanente*, 6 décembre 2010. http://www.journaldumauss.net/spip.php?article743. Accessed 15 Oct 2011.

Lorica, Ben. 2009. *Big data and real-time structured data analytics*. www.radar.oreilly.com/2009/08/big-data-and-real-time-structured-data-analytics.html. Accessed 14 Dec 2011.

Simondon, Gilbert. 1980. *On the mode of existence of technical objects*. Ontario: University of Western Ontario.

Simondon, Gilbert. 2005. *L'invention dans les techniques: Cours et conferences*. Paris: Seuil.

Stiegler, Bernard. 2010. *For a new critique of political economy*. Cambridge: Polity.

Stiegler, Bernard, and Irit Rogoff. 2010. Transindividuation. http://www.e-flux.com/journal/view/121. Accessed 17 May 2012.

Vandenberghe, Frédéric. 2001. From structuralism to culturalism: Ernst Cassirer's philosophy of symbolic forms. *European Journal of Social Theory* 4(4): 479–497.

Chapter 7
Against Environmental Protection? Ecological Modernization as "Technician Ecology"

Isabelle Lamaud

Jacques Ellul is often considered a forerunner of ecological thought, since as early as the 1930s numerous related themes began to appear in his thought: critique of the State and technician-dominated bureaucracy; the limitations of industrial society, technoscientific expansion, and capitalist organization of labor; and praise for an ascetic lifestyle and fight against the destruction of nature (Troude-Chastenet 1998). Several texts can be found specifically dealing with environmental issues (Ellul 1972, 1973) in his work, and many of his writings revolve around what would now be seen as ecological issues (notably his books on *La Technique*, Ellul 1990 [1954], 2004 [1977], 1988). He often refers to the writings of his friend and fellow thinker Bernard Charbonneau, who influenced him to take nature into account (see, for example, Charbonneau 2002 [1969], 2009 [1980]; Cérézuelle 2006), and his personal commitments found an ideal breeding ground in ecologists' struggles (in particular, the fight against the development of the Aquitaine Coast).

Nevertheless, the issue of the relationship between humanity and nature is not especially developed in his thought and remains rooted in traditional oppositions such as humanity versus nature and politics versus science, which have since been widely challenged by ecologists' thinking. Therefore I propose here to return to the work of Ellul not for his direct contribution to the understanding of the relationship between humanity and nature, but because of the relevance of his analysis of the *société technicienne* (referred to here as "technician society") in the understanding of the current modes of management of the ecological issues.

Indeed, I suggest that ecological issues are at present mainly perceived through the ideology of "ecological modernization" (EM), an ideology anchored in practices that construct a certain mode of management of environmental issues. This ideology has been the subject of a significant range of academic literature in English (for an overview of the most significant works see: Weale 1992; Hajer 1995, 1996;

I. Lamaud (✉)
CMRP, Université Montesquieu Bordeaux IV, Avenue Léon Duguit, 33608 Pessac, France
e-mail: isalamaud@gmail.com

H.M. Jerónimo et al. (eds.), *Jacques Ellul and the Technological Society in the 21st Century*, Philosophy of Engineering and Technology 13,
DOI 10.1007/978-94-007-6658-7_7,

Spaargaren et al. 2000; Mol and Spaargaren 2000; Mol 2001; Dryzek 2005; Mol et al. 2009), characterized by a diversity and a complexity reflecting the flexibility of the idea of ecological modernization. I propose here to see how taking into account the technical factor as a separate issue, as suggested by Ellul, can enlighten challenges raised by ecological modernization and helps to make better sense of the fundamentally technician[1] character of this approach.

First I will give a brief overview of the mode of conceptualisation dominating environmental management through the ideology of ecological modernization. Then I will see how isolating the technical factor leads to better understanding and makes it possible to go beyond the limits of an economic critique. Finally I will show, based on the Ellulian perspective, how ecological modernization can be considered as the continuation of technical development through the reduction of the ecological issue to "environmental protection."

1 The Ideology of Ecological Modernization: Restructuring Modern Societies Based on Environmental Rationality

1.1 The Ideology of Ecological Modernization as a Conceptual Framework

The term "ecological modernization" has encountered significant success in literature in German and English. Ideas attached to this term can also be found (at least partly) in the literature revolving around other terms: sustainable development, industrial ecology, environmental market liberalism, etc. I choose here to refer to the ideology of ecological modernization since:

- Several distinctive approaches of environmental management appear to share common characteristics linked to the general context of modernity which are explicit in the literature on ecological modernization.
- The literature concerning ecological modernization seems at the same time more precise than the literature concerning sustainable development and more exhaustive than specific approaches such as industrial ecology and environmental market liberalism.
- The theorising research work done on ecological modernization, particularly by the research team of Mol and Spaargaren in Wageningen, the Netherlands, provides solid bases for analyses.

I will present the ideology of ecological modernization only briefly. It is a complex object of research, insofar as its meaning has changed over time. Since its emergence at the beginning of the 1980s, it has taken on different meanings depending on who

[1]The term "technician" will be used to refer to the Ellulian concept as in *la société technicienne* or *le système technicien*.

uses it, and has referred both to a process of social change (which takes place on its own and is only observed) and to a discourse conveyed by different social players who intend to promote it. This last element justifies our choice to consider ecological modernization as an ideology: indeed, according to Raymond Aron, "political ideologies always combine, more or less felicitously, factual propositions and value judgements"[2] (Aron 2002 [1955]: 246). The confusion between descriptions and prescriptions is a constant characteristic of the literature on ecological modernization, and particularly of the researchers who have been working on its development under the form of a social theory (mainly Mol and Spaargaren). In short, the ideology of ecological modernization refers here to a system of beliefs and values attached to a project of environmental restructuring of modern societies. It proposes an interpretation of the environmental crisis, which differentiates itself both from the position of denial of this crisis and from its interpretation by radical ecology. Ecological modernization suggests that modern societies have entered a process of environmental reform. This is characterized by the emergence of an ecological rationality which is supported by technological developments and market-based instruments that exist within the framework of an integrated management of the environment by a coalition of diverse social players, anchored in modernist beliefs and a positivist epistemology.

1.2 The First Wave in the Construction of Ecological Modernization: Market Mechanisms and Technological Development

When ecological modernization emerges at the beginning of 1980s, it marks the start of a serious acknowledgement of the ecological issue by diverse social players in the context of the increasing impact and complexity of environmental problems (acid rain, loss of biodiversity, climate change, etc.). The ecological crisis is slowly recognized as a unique situation calling for a deep restructuring of so-called modern societies. Among certain international organizations (particularly the OECD) and academic circles (e.g., Huber and Jänicke in Germany), the links between environmental and economic issues are reconceptualized so as to make environmental protection and economic development compatible. Faced with the questioning of economic growth by radical ecology, ecological modernization reaffirms that it is possible, and even necessary, to organize environmental protection and economic development so that they mutually reinforce each other, making the involvement of powerful business players in the protection of the environment possible.

In this stage of ecological modernization, the environmental crisis is considered as a problem to be regulated by market forces, which have heretofore failed to take

---

[2]"Les idéologies politiques mêlent toujours, avec plus ou moins de bonheur, des propositions de fait et des jugements de valeur" (our translation).

negative environmental externalities into account. This problem must be tackled by internalizing environmental costs, which means first quantifying them in order to integrate them in the market. This is economically legitimized by the idea that "pollution prevention pays" for the costs of cleaning up environmental degradation can turn out to be higher than the cost of its prevention. The (re)conceptualization of the environmental crisis within the economic framework ("ecologizing economy," "economicizing ecology") relies on the idea that pollution is a sign of waste and inefficiency, therefore damaging both the environment and the economy. Industrial production should therefore be restructured along environmental lines (see, for example Mol 1995): re-using waste, increasing eco-efficiency (producing more goods and services by using fewer resources and producing less waste and pollution), etc. The State can also find a comparative advantage here benefiting national economic growth and employment. The relevance of ecological modernization would be grounded by cases of OECD countries that managed to "decouple" their economic growth from the growth of their resource use between the 1970s and 1990s (the debates around this assertion will not be developed here).

In a time of general questioning of the role of the State by neo-liberal critiques, state management of the environment is itself considerably criticized (and stigmatised as "command-and-control"; cf. Weale 1992). German researchers who studied and promoted ecological modernization in the 1980s saw a central role for technological development and market mechanisms. The State must simply create the conditions in favor of the pursuit of technological innovation by Schumpeterian entrepreneurs. Along evolutionist lines, Huber identified several stages in the development of industrial societies: the first one, up to 1948, characterized by the breakthrough of industrialisation, the second one up to 1980 by its construction, and the third one by the ecological modernization of industrial development. According to Huber (quoted in Mol 1995: 35–38), the stage of ecological modernization is characterized by "superindustrialisation," where more industrialisation and modernization are necessary in order to overcome environmental problems. I find this type of deterministic approach in the current reappropriation of the idea of ecological modernization by Chinese researchers (Mol 2006; Research Group for China Modernization Strategies 2007). I will see that this approach has been criticized and made more complex right at the very heart of ecological modernization; nevertheless, the basic premises remain: a central role for technological innovation and market regulation, a modernist approach anchored in a progressive philosophy of history.

1.3 The Second Wave of the Construction of Ecological Modernization: Environmental Rationalization Through the Mobilization of All Social Players

In this stage, the evolutionist and deterministic approach of ecological modernization is toned down, with Mol and Spaargaren criticising the technological optimism of Huber and preferring the idea that technique constitutes part of the problem as much as part of the solution. Mol and Spaargaren anchor ecological modernization more deeply within social theories of modernization by defining it as the rationalisation of production and consumption processes along environmental lines. They suggest that the emergence of a new sphere of environmental rationality can be observed, leading modern societies toward a new stage of their development (a process qualified by Mol as the "emancipation of ecology"). According to them, modern societies have entered this process of environmental reform with results that are significant enough to invalidate the hypothesis of the radical ecology movement: that the bases of modernity should be questioned in order to tackle the ecological crisis.

> EMT [Ecological Modernization Theory] challenged the environmental movement's traditional idea that a fundamental reorganization of the core institutions of modern society (the industrialized production system, the capitalist organization of the economy and the centralized state) was essential in entering a path of long term sustainable development (Mol and Spaargaren 2000: 19).

They specify the fact that this process is not automatic, and that one should build up appropriate governance by mobilizing the main social players identified (a socio-political change which has been termed "political modernization" [Jänicke 2009; Tatenhove and Leroy 2003]):

- *The State*, which must favor market mechanisms, and integrate the environmental criteria in all its policies.
- *The business sector* at the heart of the ecological modernization processes. Following the internalization of environmental costs and the competition for green technological innovation, it will be in *producers*' best interest to ecologically rationalize their production, and this even more since it will match *consumer* demand.
- Moderate *environmental organizations* (i.e., those whose claims are compatible with the current political, economic, social and technical system) play an essential role of expertise made possible by their professionalisation and institutionalization in the 1980s.
- *Scientists* who must produce the knowledge and techniques required for ecological modernization (for example, researchers who analyze ecological modernization serving as advisers for diverse social players). Science and technique have an essential role that should be preventive rather than curative.

This overview of the development of ecological modernization ideology shows its significant flexibility: it can adapt just as well to a neo-liberal context as to a more State-centered type of environmental management.

2 An Ellulian Approach: The Relevance of the Technical Factor

If economic and technical growth is indeed interwoven in the complexity of the real world, here I suggest following Ellul's methodology in theoretically distinguishing (as ideal types) different factors determining social order, in particular the economic and the technical factor:

> Confusion between Technique and Science, between Technique and Machine, [...] even more frequent confusion between Technique and Economy. As soon as an attempt at dissociation is made, Marxists make accusations that diversionary tactics are being used

> and an idealistic antirevolutionary attitude adopted! And yet, as long as we haven't studied the technical phenomenon beyond its economic implications and beyond the issues of the economic system and class struggle, we condemn ourselves to not understanding anything about contemporary society (and therefore not being able to carry out any revolutionary action either!)[3] (Ellul 2004 [1977]: 39–40).

This distinction seems necessary so as to understand ecological modernization in its complexity and not to reduce its meaning to capitalistic greenwashing.

2.1 The Reduction of Ecological Modernization to the Free Market Side of Sustainable Development

Analyses often tend to focus on the economic factor in order to make sense of ecological modernization. Thus EM is presented as the ideology of green (economic) capitalism, of green (economic) growth. It is then distinguished from the idea of sustainable development by the preponderance given to business players and to market mechanisms (Zaccaï 2002; Bäckstrand and Lövbrand 2007; Carter 2007), and by the lack of consideration given to intra- and intergenerational social justice demands (Langhelle 2000). These types of analysis tend to associate ecological modernization with a mode of free market management meeting private companies' economic interests.

According to our analysis, ecological modernization can nevertheless not be reduced to private sector interests or a free market or neo-liberal ideology. As Murphy puts it, it seems to be a common misinterpretation to conceive ecological modernization as an approach to environmental problems based on market liberalism (Murphy 2000: 1). Indeed the countries which have implemented it most successfully were identified as neo-corporatist states in the Dryzek et al. study (Germany, the Netherlands, Sweden, Norway, Japan, etc.), whereas states sticking closer to market liberalism such as the United Kingdom or the United States have more difficulty in following the path of ecological modernization (Dryzek et al. 2003). According to Dryzek, ecological modernization lends itself even less than sustainable development to market liberalization:

> I argued [...] that sustainable development fits uneasily in a world seemingly committed to free trade and the deregulation of markets [...]. Ecological modernization faces even greater problems here, given its commitment to conscious collective control of the political economy in the ecological restructuring of capitalism. [...] Concerted pursuit of ecological

[3]My translation from the French: Confusion entre Technique et Science, entre Technique et Machine, [...] confusion encore plus fréquente entre Technique et Economie. Sitôt que l'on essaie de dissocier, les marxistes accusent de faire une manœuvre de diversion et d'avoir une attitude idéaliste antirévolutionnaire! Et pourtant, aussi longtemps que l'on n'aura pas étudié le phénomène technique en dehors de ses implications économiques et des problèmes de système économique ou de lutte de classe, on se condamne à ne rien comprendre de la société contemporaine (et par conséquent à ne rien pouvoir faire non plus comme action révolutionnaire!).

modernization requires a consensual and interventionist policy style consistent with corporatism (Dryzek 2005: 177).

Mol and Spaargaren (2000) themselves insist on the fact that it is neither a matter of free market nor mere internalization of environmental costs as traditional neoclassical economics would advocate. Indeed ecological modernization has turned out, as seen in the first part of this paper, to be flexible on the issue of market mechanisms and business players, and the role of the State has gained central importance ever since the 1990s.

2.2 *Marxist-Inspired Economic Analyses: Technique as an Unexamined Factor*

Critical standpoints within academic debates on ecological modernization can be divided into two main groups: internal critiques which do not question the epistemological and ideological premises of ecological modernization, suggesting merely a more in-depth study of its conditions of implementation, and external critiques dealing with its epistemological and ideological premises. Among the latter, the dominant approach may be seen as Marxist-inspired and based on economic issues. It focuses on the incompatibility between economic development along capitalist lines and environmental protection. US environmental sociology, as exemplified by O'Connor, Schnaiberg, and Goldblatt, has largely represented this current, which Mol and Spaargaren (2000) call "neo-Marxist." The relevance of this kind of analysis is that it makes clear the conflicts among differing economic interests and between dominant economic interests – and a system based on infinite economic growth – and the protection of the environment. It has been helpful in raising the issue of social justice in the management of environmental issues and in questioning the role of business players. It has questioned the first wave of ecological modernization and its belief in the capacity of the market to take care of environmental externalities. It has questioned technological optimism in a context where technical innovation remains dominated by private companies' interests. It has also highlighted the fact that polluting industries may merely have moved to third-world countries while continuing to market to first-world consumers. Yet this critique tends to ignore issues raised by ecological modernization outside of the economic sphere. By reducing its critique to (economic) capitalist aspects of ecological modernization, it leaves the technical factor unexamined. The mechanical, economic, organizational and human techniques (according to Ellul's typology, 2004 [1977]) advocated by the ideology of ecological modernization are then criticized only to the extent that they serve capitalist economic interests, which amounts to maintaining the belief in the neutrality of techniques themselves. The challenge is reduced to the question of whether a technique is in hands that will determine the positive or negative character of its use. Yet the belief in the efficiency of ecological modernization's ideas and processes in facing environmental problems seems to us to go far beyond economic interests. This limit to the Marxist-inspired critique might explain why Mol and Spaargaren (2000) can claim that this type of critique can be (and has been) successfully integrated by ecological modernization.

2.3 *The Flexibility of Ecological Modernization and Its Technical Premises*

Rather than focusing on the capitalist aspects of ecological modernization, I propose to follow Ellul's methodology and to consider the technical premises of ecological modernization. The ideology presents mechanical, economic, organizational and human techniques as neutral and efficiency-based management strategies for tackling the environmental crisis. Our analysis of ecological modernization above suggests that it can take various forms, relying more or less on market-based instruments, strong state intervention, as well as the rationalisation of consumers and citizens' behaviors or large-scale technological innovations. The constant factor remains the belief in the capacity of a technician form of management in the face of environmental problems.

The current of ecological modernization presents itself as a neutral, non-ideological approach, merely intending to resolve a technical problem in the most efficient way (Blühdorn 2000). The emergence of an independent sphere of environmental rationality makes the technoscientific management of natural resources possible (according to Spaargaren, it is a matter of developing a "'hard' science of the sustenance base" [Spaargaren et al. 2000: 51]) by disembedding the management of the environment from other issues:

> The point of reference for this radical transformation is the movement toward an environmentally sound society, and not a variety of other social criteria and goals, such as the scale of production, the capitalistic mode of production, workers' influence, equal allocation of economic goods, gender criteria and so on (Mol 1996: 309–310).

According to Mol it is a matter of disembedding the environmental sphere from the "socio-ideological" sphere. This process of rationalization of environmental management implies, on the one hand, that the ecological issue be reduced to one of natural resources management, entailing a certain vision of nature as a separate entity that could then be objectively studied and scientifically controlled; on the other hand, the idea of a rationality detached from socio-ideological factors implies a neutral and naturalizing type of management by experts founded on an unquestionable criterion of efficiency.

The technical variable is thus central to the ideology of ecological modernization, as has been noted by several researchers. According to Hajer, "conceptually EM relies heavily on science, technology, and expert-led processes of change," and it is characterized by "a renewed belief in the possibility of mastery and control" (Hajer 1995: 35). "Denied are notions that nature might spring surprises on us, defy human management" (Dryzek 2005: 170). Thus although Mol distances himself from the technological deterministic approach of Huber, he nevertheless leaves the technical factor unexamined under the guise of its neutrality (with the idea that technique can at the same time be part of the problem and of the solution). Other researchers have more clearly stated the role technique could or should play. Cohen highlights in several of his articles that the capacity of nation-states to adopt ecological modernization (which he identifies as "the new voice of rational ecology") depends on their propensity to align their policies to "strict rational reasoning" (Cohen 1998, 2000). He affirms, "[T]his policy programme is dependent upon a firm commitment to science and a preference to address environmental problems in technological terms" (Cohen 2000: 77). Ecological modernization can then be interpreted as embedded within the modern framework as studied by Latour (1997), where science and technique are strictly separated from social and political concerns, with their mode of development being neutralized and no longer challenged. By proposing that environmental problems be treated as objects with technical solutions, ecological modernization discourages the questioning of modernist beliefs sustaining the myth of technical progress.

3 Ellul Against "Environmental Protection": The Technicization of Environmental Management

I propose here to focus on the fact that ecological modernization can be analyzed as an ideology of which a main characteristic is to allow the pursuit of technical development and the preservation of the belief in technical management in the face of the ecological movement which has questioned them. In this perspective, this ideology then constitutes a fundamental element in the extension of the ascendancy of technique and technicians analyzed by Ellul.

3.1 Ecology Against "Environmental Protection"

In his "plea against environmental protection" ("Plaidoyer contre la défense de l'environnement" published in 1972), Ellul explains that his purpose is not to question the fight against the destruction of nature, which he has supported since the 1930s, but to condemn the contradictions of "environmental protection" insofar as it develops in parallel with technical growth. Thus,

> taking an interest in the protection of the environment and in ecology without questioning technical progress, the technician society, the passion for efficiency, is to undertake an operation which is not only useless, but fundamentally harmful. Because it will lead to nothing, but one will have the impression of having done something, it will allow one to falsely calm legitimate worries by throwing a new shroud of propaganda over the threatening reality[4] (Ellul 1972: 11).

[4]"S'intéresser à la protection de l'environnement et à l'écologie sans mettre en question le progrès technique, la société technicienne, la passion de l'efficacité, c'est engager une opération non seulement inutile, mais fondamentalement nocive. Car elle n'aboutira finalement à rien, mais on aura eu l'impression d'avoir fait quelque chose, elle permettra de calmer faussement des inquiétudes légitimes en jetant un nouveau voile de propagande sur le réel menaçant" (my translation).

If ecology is about the relationship of human beings with nature, or about how human beings handle their means of existence, and technique constitutes a fundamental intermediary between human beings and their means of existence, then ecology calls for questioning the empowerment of technique. On one hand, the ecological movement has been an important field for challenging technical growth. Controversies linked to the use of specific techniques, such as automobiles, nuclear energy, pesticides, genetically modified organisms, and nanotechnology, lie at the heart of ecological issues. Increased consciousness of the problems with these techniques disrupts the belief in the neutrality of technical progress and the dichotomies of modern thought, such as nature versus culture, the scientific versus the social or political, facts versus values, and the objective and unquestionable versus the subjective and questionable. On the other hand, since its emergence the ecological movement has been a field for the profusion of techniques where research, experimentation, knowledge and transmission of know-how hold a central place (such as the fields of energy, health, agriculture, and housing). The ecological movement has opposed the monopolization and standardization of knowledge and know-how in the process of technical growth and the concomitant impoverishment of the relationship between human beings and techniques, in the sense that human beings no longer master them. My thesis is that by raising the possibility of "environmental protection" without questioning technical growth, ecological modernization has indeed allowed, as Ellul feared, major issues brought up by the ecological movement to be ignored.

3.2 *Technician Management of the Environment*

The questioning of the idea of technological neutrality lies at the center of Ellul's analyses:

> [T]echnique implies in itself a certain number of consequences, represents a certain structure, certain demands, generates certain modifications of humans and society, that are imposed upon us whether we like it or not... I'm not saying that this is absolutely irremediable, but rather that in order to change this structure or reorient this movement we must make a huge effort to take over what we believed to be mobile and adjustable, we need to become aware of this independence of the technician system, which is opposed by the reassuring conviction of the neutrality of technique[5] (Ellul 2004 [1977]: 162).

[5]"La technique a en soi un certain nombre de conséquences, représente une certaine structure, certaines exigences, entraine certaines modifications de l'homme et de la société, qui s'imposent qu'on le veuille ou non. [...] Je ne dis pas que c'est absolument irrémédiable, mais que pour changer cette structure ou orienter différemment ce mouvement il faut un effort immense de prise en main de ce que l'on croyait mobile et orientable, il faut la prise de conscience de cette indépendance du système technicien, à quoi s'oppose la conviction rassurante de la neutralité technique" (my translation).

Ellul is not saying that technique is either good or bad in itself, but warning us that any technical configuration contributes to shaping a society. It cannot be thought of separately from its use, and it has ambivalent consequences on relationships among human beings, their relationships to nature and the conditions of their existence and the configurations of power in that society, just as do economic, political and social factors. One must consider these implications in terms of the aims and values held and to be able to refuse specific technical configurations that seem inconsistent with these aims and values. Yet the belief in the neutrality of technique makes it possible to blindly forge ahead into unexamined technical development. Consequently technique automatically expands, is universalized and becomes autonomous from human decision-making. The technical phenomenon then fundamentally contradicts the ideal of human freedom.

Ecological modernization can be interpreted as part of the framework of the technician society:

- *It shapes a technical answer to a problem it has itself defined as being technical* ("In fact, each solution is technical, and defines itself the problem"[6] [Ellul 2004 [1977]:281]). Indeed, isolating an environmental sphere makes its objectification and its quantification possible, and then its management by environmental experts necessary, since they will be the most qualified in these conditions to protect the environment most efficiently.
- *It maintains the belief in the neutrality of technique*: the management tools proposed are presented as fulfilling first and foremost the criteria of efficiency. Market mechanisms and state regulation can be used equally depending on their efficiency. Technical developments can be positive or negative: they may have contributed to the emergence of the environmental crisis, but they will also be able to contribute to its resolution, depending on how they are used.
- *It maintains the belief in the ineluctability of technical development*: the transformations advocated by ecological modernization are seen as representing the only way to manage the environmental crisis, and they are already occurring and not to be questioned.

3.3 "Environmental Protection" as a Legitimizing Factor of Technical Growth

Thus I see that environmental problems have provided a major field of legitimization for the growth of the ascendancy of technique, something Ellul had not foreseen in this specific field but which corresponds to his analyses concerning technical growth. Mechanical, economic, organizational and human techniques are brought

[6] "En réalité, chaque solution est technique, et définit elle-même le problème" (my translation).

together in the fight for environmental protection, expanding to new spheres of human life and new geographic areas. These developments confirm Ellul's thesis, according to which, destruction caused by the technician system does not constitute a barrier to its further growth, but on the contrary can create new opportunities for expansion: "The great mechanism of production is self-augmentation, which is in reality the emergence of problems, dangers and difficulties"[7] (Ellul 2004 [1977]: 231). When already existing techniques are imposed under the guise of environmental protection, finding a goal only in hindsight, Ellul's analyses concerning the autonomous development of technique are confirmed: "Technique does not develop according to specific ends to pursue but according to already existing possibilities of growth"[8] (Ellul 2004 [1977]: 263). Finally, in the face of technical development, the expansion of technical ascendancy appears as the only remaining possibility. For example, it no longer seems possible to manage the nuclear industry without calling on experts and sophisticated techniques, which will have to manage the situation in the long run.

Isolating the technical factor at a theoretical level, while keeping in mind the idea that it is actually inseparable from other factors, makes it possible to further the analysis of ecological modernization. Environmental management seen from an Ellulian perspective presents ecological modernization as an ideology that fails to question technical progress. From there, it can be put in opposition with the idea of a philosophy of ecology that does question technical progress. Such a philosophy would not be technophobic, refusing any technique whatsoever, but would rather make the effort to understand technique and make visible its political consequences, by contrast with the dominant trend of evicting the issue of technical autonomy from socio-political thinking (Winner 1977). Such questioning holds the potential for curbing the autonomy of technique, because, as Ellul claims throughout his work, it is only when human beings give up questioning technical development that it becomes autonomous.

References

Aron, Raymond. 2002 [1955]. *L'opium des intellectuels*. Paris: Hachette.

Bäckstrand, Karin, and Eva Lövbrand. 2007. Climate governance beyond 2012: Competing discourses of green governmentality, ecological modernization and civic environmentalism. In *The social construction of climate change: Power, knowledge, norms, discourses*, ed. Mary E. Pettenger, 123–148. London: Ashgate.

[7]"Le grand mécanisme de production de l'auto-accroissement, c'est en réalité l'apparition des problèmes, dangers et difficultés" (my translation).

[8]"La technique ne se développe pas en fonction de fins à poursuivre mais en fonction des possibilités déjà existantes de croissance" (my translation).

Blühdorn, Ingolfur. 2000. Ecological modernization and post-ecologist politics. In *Environment and global modernity*, ed. Gert Spaargaren, Arthur P.J. Mol, and Frederick H. Buttel, 209–228. London: Sage Publications.
Carter, Neil. 2007. *The politics of the environment: Ideas, activism, policy*, 2nd ed. New York: Cambridge University Press.
Cérézuelle, Daniel. 2006. *Ecologie et liberté: Bernard Charbonneau, précurseur de l'écologie politique*. Lyon: Parangon.
Charbonneau, Bernard. 2002 [1969]. *Le jardin de Babylone*. Paris: Editions de l'Encyclopédie des Nuisances.
Charbonneau, Bernard. 2009 [1980]. *Le feu vert: Autocritique du mouvement écologique*. Lyon: Parangon.
Cohen, Maurie J. 1998. Science and the environment: Assessing cultural capacity for ecological modernization. *Public Understanding of Science* 7: 149–167.
Cohen, Maurie J. 2000. Ecological modernisation, environmental knowledge and national character: A preliminary analysis of the Netherlands. *Environmental Politics* 9(1): 77–106.
Dryzek, John S. 2005. *The politics of the earth: Environmental discourses*. Oxford: Oxford University Press.
Dryzek, John S., David Downes, Christian Hunold, et al. 2003. *Green states and social movements: Environmentalism in the United States, United Kingdom, Germany, and Norway*. Oxford: Oxford University Press.
Ellul, Jacques. 1972. Plaidoyer contre la "défense de l'environnement." *France Catholique*, n.° 1309, 1310, 1311. Brochure created and published by *Notes & Morceaux Choisis*.
Ellul, Jacques. 1973. Le mythe de l'environnement. *Economies et sociétés*: *Cahiers de l'I.S.E.A*, séries F. 25 7(9):1539–1554.
Ellul, Jacques. 1988. *Le bluff technologique*. Paris: Hachette.
Ellul, Jacques. 1990 [1954]. *La technique ou l'enjeu du siècle*. Paris: Economica.
Ellul, Jacques. 2004 [1977]. *Le système technicien*. Paris: Calman-Lévy, Le Cherche Midi.
Hajer, Marteen A. 1995. *The politics of environmental discourse: Ecological modernization and the policy process*. Oxford: Clarendon Press.
Hajer, Marteen A. 1996. Ecological modernization as cultural politics. In *Risk, environment and modernity. Toward a new ecology*, ed. Scott Lash, Bronislaw Szerszynski, and Brian Wynne, 246–268. London: Sage.
Jänicke, Martin. 2009. On ecological and political modernization. In *The ecological modernization reader: Environmental reform in theory and practice*, ed. Arthur P.J. Mol, David A. Sonnenfeld, and Gert Spaargaren, 28–41. London/New York: Routledge.
Langhelle, Oluf. 2000. Why ecological modernization and sustainable development should not be conflated. *Journal of Environmental Policy & Planning* 2(4): 303–322.
Latour, Bruno. 1997. *Nous n'avons jamais été modernes: essai d'anthropologie symétrique*. Paris: La Découverte.
Mol, Arthur P.J. 1995. *The refinement of production: Ecological modernization theory and the chemical industry*. Utrecht: Jan van Arkel/International Books.
Mol, Arthur P.J. 1996. Ecological modernisation and institutional reflexivity: Environmental reform in the late modern age. *Environmental Politics* 5(2): 302–323.
Mol, Arthur P.J. 2001. *Globalization and environmental reform: The ecological modernization of the global economy*. Cambridge, MA: MIT Press.
Mol, Arthur P.J. 2006. Environment and modernity in transitional China: Frontiers of ecological modernization. *Development Change* 37(1): 29–56.
Mol, Arthur P.J., and Gert Spaargaren. 2000. Ecological modernisation theory in debate: A review. *Environmental Politics* 9(1): 17–49.
Mol, Arthur P.J., David A. Sonnenfeld, and Gert Spaargaren (eds.). 2009. *The ecological modernization reader: Environmental reform in theory and practice*. London/New York: Routledge.
Murphy, Joseph. 2000. Ecological modernisation. *Geoforum* 31: 1–8.

Research Group for China Modernization Strategies. 2007. *China modernization report: Ecological modernization*. Beijing: Peking University Press.

Spaargaren, Gert, Arthur P.J. Mol, and Frederick H. Buttel. 2000. *Environment and global modernity*. London/Thousand Oaks: Sage Publications.

Troude-Chastenet, Patrick. 1998. Jacques Ellul, précurseur de l'écologie politique? *Ecologie et Politique* 22: 105–119.

van Tatenhove, Jan P.M., and Pieter Leroy. 2003. Environment and participation in a context of political modernisation. *Environmental Values* 12: 155–174.

Weale, Albert. 1992. *The new politics of pollution*. Manchester: Manchester University Press.

Winner, Langdon. 1977. *Autonomous technology: Technics-out-of-control as a theme in political thought*. Cambridge/London: MIT Press.

Zaccaï, Edwin. 2002. *Le développement durable: Dynamique et constitution d'un projet*. Bruxelles: Peter Lang.

Part II
Autonomous Technology

Chapter 8
Propaganda and Dissociation from Truth

Langdon Winner

Jacques Ellul's *Propaganda* is a work of analysis and explanation about a crucial range of practices and institutions in modern political society. Drawing upon the social history of political communication and a wealth of social scientific studies on public opinion published during the middle twentieth century, Ellul expands the framework in which such evidence can been understood and put to use. Especially at the book's conclusion, it is clear that Ellul hoped the book would serve as a warning to democratic citizens of technological societies about the insidious spread of propaganda throughout the body politic.

But astute descriptions about dangers in the world can, in the hands of malevolent actors, become guidebooks for how to produce the maladies in question. The very same arguments and observations meant to urge caution can just as well be used as a guide, a how-to-do-it book for those who want to practice this highly toxic art. While it is unlikely that the corporate executives, politicians and media specialists of our time have actually been reading *Propaganda*, it appears that some of its most riveting insights – perhaps derived from other sources – have now become central features in the playbook of the most prominent propaganda mills to have emerged in the United States in recent times. Within the daily fare of distorted "news" broadcast to millions of viewers on cable and satellite television channel are vivid examples of a malevolent flight from reality equal to the nightmares in both Ellul's and George Orwell's worst imaginings. If nothing else, Ellul's book can be read as a premonition of the awful predicament that even relatively stable and prosperous democracies now face.

L. Winner (✉)
Department of Science and Technology Studies, Rensselaer Polytechnic Institute,
Troy, NY 12180-3590, USA
e-mail: winner@rpi.edu

H.M. Jerónimo et al. (eds.), *Jacques Ellul and the Technological Society in the 21st Century*, Philosophy of Engineering and Technology 13,
DOI 10.1007/978-94-007-6658-7_8, © Springer Science+Business Media Dordrecht 2013

1 Modern Society's Need for Propaganda

In Ellul's account, "propaganda" is the name for a range of techniques that characterize communication within the large, technology-centered societies of the modern world. Seeking a comprehensive definition of these practices, he probes a range of concepts from the conventional literature on mass communications and eventually offers a serviceable definition in two parts. "Propaganda is a set of methods employed by an organized group that wants to bring about the active or passive participation in its actions of a mass of individuals, psychologically unified through psychological manipulations and incorporated in an organization" (Ellul 1965: 61). In a later paragraph Ellul elaborates further, noting that propaganda is "the group of manifestations by which any society seeks to integrate the maximum number of individuals into itself, to unify its member's behavior according to a pattern, to spread its style of life abroad and thus to impose itself on other groups" (Ellul 1965: 62). From Ellul's point of view, propaganda is an aggressive form of thought and communication, one whose underlying tendency is to establish dominance over other available constructions of social reality. The ideas of a particular group, institution, corporation, political party, or nation state define a way of life that its adherents strongly believe is preferable to all others. Thus, propaganda is closely linked to modern political ideologies, ones often spread with missionary zeal. People who embrace a particular ideology believe that their way of being is clearly superior to all others and that those outside the group should be persuaded to accept that conviction as well.

Much of Ellul's attention focuses upon the specific methods that enable propagandists to shape people's perceptions, opinions, and sentiments. Among the familiar techniques he mentions are: selective rendering of facts; deliberate use of falsehoods; spreading of rumors; descriptions of groups and individuals through stereotyping and innuendo; the invocation of prejudice, fear, and hatred within the populace; and the incessant repetition of emotionally laden cultural myths and divisive slogans. Methods of this kind, he argues, are pervasive in modern society, and are employed to some extent by a great many groups who wish to shape public opinion in their favor.

Despite the intrinsic fascination that particular practices hold for observers of mass communications, it is misleading, Ellul contends, to emphasize the particular strands of persuasion and the skillfully crafted techniques by which they are deployed. His central argument is that propaganda must be seen as a highly general social phenomenon, one whose specific practices matter less than the continuing, powerful presence of propaganda in our lives, a pervasive way of creating and transmitting messages that affect everyone, everyday within a mass populace. Far more than a collection of ingenious means to persuade people to adopt one set of beliefs rather than another, propaganda has become something like the collective, central nervous system of complex, industrialized, bureaucratized, media-centered societies.

> The more the techniques of distributing information develop, the more the individual is shaped by such information. It is not true that he can choose freely with regard to what is

presented to him as the truth. And because rational propaganda thus creates an irrational situation, it remains, above all, propaganda – that is, inner control over the individual by a social force, which means that it deprives him of himself (Ellul 1965: 87).

The conditions that make it possible for the distortions of propaganda to spread and flourish, in Ellul's view, stem from a series of crises within the modernity – the economic, social, and political upheavals of the eighteenth and nineteenth centuries – that have fragmented society and weakened people's ability to find coherence and meaning in the churches, workplaces, and local communities that formerly sustained them. Faced with the complicated, confusing, often distressing situations that modern, technological societies present, propaganda offers ways of understanding that, for better or worse, give a pattern to events and help largely isolated individuals find meaning within the perplexing, artificially structured world that confronts them.

An individual can be influenced by forces such as propaganda only when he is cut off from membership in local groups. Because such groups are organic and have a well-structured material, spiritual, and emotional life, they are not easily penetrated by propaganda (Ellul 1965: 91).

Ellul's argument here is resonant with those "mass society" theories advanced by political sociologists in the decades after World War II. In attempts to explain the rise of Fascism, Nazism, and Communism, writers such as William Kornhauser (1959) pointed to the disintegration of social buffers that previously stood between individuals and the leaders who held power in social movements, political parties, and the centralized state. Within distressed economies and fragmented social systems, large numbers of people became available for mobilization within what would become revolutionary movements and, eventually, the totalitarian regimes these movements tended to produce. While emphasizing the formative influence of propaganda within the toxic brew, Ellul's book comments extensively on social and political developments of this kind.

As Ellul explores the techniques of mass communications, he introduces a number of sociological distinctions that help illuminate the multiplicity of ways in which propaganda can be effective – vertical vs. horizontal and disruptive vs. integrative, among others. Thus, in the mode of vertical communication, sources at the top of the institutional hierarchies craft messages that are broadcast to those at lower levels of the pyramid of social influence. Of course, this has become the dominant pattern in state and corporate controlled media of the twentieth century. In contrast are horizontal methods in which people themselves at middle and lower-levels of society are mobilized to carry the message – an example of which can be found in the nationwide grassroots propaganda campaigns in China instigated by Mao Zedong. While the messages originated from the Party and its revolutionary leader, their propagation moved from person to person, group to group, in mass, grassroots mobilizations; the rituals of "the little red book," for example. Another key contrast is that between disruptive and integrative approaches. Propaganda campaigns of disruption are ones that emphasize sources of injustice, conflict, and disorder in political society in the hope of gaining advantage, perhaps revolutionary leverage, from a growing awareness of such troubles within the populace that receives the message. Propagandas of integration are ones that stress features of everyday life exhibiting harmony, order, and connection, while associating such qualities with particular leaders or organizations. A good illustration of a successful theme of social integration would be Ronald Reagan's "It's morning again in America" television commercial of 1984, which showed happy, smiling people in the towns and villages of a contented nation.[1]

Although distortion and thinly concealed lying are widely recognized features of the propagandist art, Ellul insists that all successful messaging must contain a strong supply of verifiable facts. The confusion and discomfort of an audience can be relieved and manipulated by providing a package of carefully selected facts within a framework of meaning that seems to restore order and provide relief. Indeed, it is an oversupply of factual information that often proves most captivating. "A surfeit of data," he writes,

> far from permitting people to make judgments and form opinions and form ends, prevents them from doing so and actually paralyzes them. They are caught in a web of facts and must remain at the level of the facts they have been given (Ellul 1965: 87).

My brief summary of his theory may reinforce the view that Ellul believes people in modern society to be passive subjects, mere victims of the skillful practitioners of the art of propaganda. His actual position takes his readers in a much different direction, although by no means a more salutary one. *Propaganda* goes far beyond other writers on the topic to argue that propaganda must be seen as something that people need and desperately desire in much the way that a drug addict needs a fix. Thus, the consumer of propaganda is by no means an innocent receptacle, but rather a participant who seeks and even provokes the psychological action of propaganda, a person who willingly lends himself to its enticements and derives considerable satisfaction from them. Critics of Ellul's book, especially media theorists of more recent times, sometimes fault him for not noticing the ways in which members of the listening and viewing audience "appropriate" electronic media for their own purposes and are thereby released from its corrupting influence. But agree with him or not, Ellul is clearly aware of such possibilities and passes stern judgment upon them. He calls attention to the view that there is an "aggressive and totalitarian political machine which pounces on the innocent victim – the individual" (Ellul 1965: 138). Of course, from this standpoint, "[T]he individual then appears helpless and crushed by gigantic forces" (Ellul 1965: 138). In response he argues:

> But I think that propaganda fills a need of modern man, a need that creates in him an unconscious desire for propaganda… Naturally, he does not say: 'I want propaganda.' On the contrary, in line with preconceived notions, he abhors propaganda and considers himself a 'free and mature' person. But in reality he calls for and desires propaganda that will permit him to ward off certain attacks and reduce certain tensions (Ellul 1965: 138).

Thus, the hope that many enthusiasts profess for the ability of consumers and young professionals to adapt communication technologies (along with the organizations

[1]Ronald Reagan TV ad: "It's morning in America again," YouTube. http://www.youtube.com/watch?v=EU-IBF8nwSY. Accessed 1 November 2011.

that manage them) to more creative, liberatory purposes, is one that Ellul regards as delusional. Writing today, in the era of the smartphone, he might say that there is no clever "app" to resolve the dilemmas posed by electronic media.

2 We Report, You Decide

The contemporary example of technically sophisticated and politically potent propaganda I will probe here is that generated 24 hours a day by the wildly popular American television channel, Fox News. While the corporation draws upon the classic moves and methods perfected by earlier masters of media manipulation, it carries them to new peaks of technical and artistic sophistication, achieving extraordinary levels of success.

Fox News Channel was founded in 1996 by Australian newspaper magnate Rupert Murdock, notorious for the use of gossip and tawdry, heavily sexualized content in his tabloid papers, the *News of the World* in London and the *New York Post*. To guide the fledgling enterprise, Murdock hired Roger Ailes, a media specialist who first came to prominence as political campaign advisor to Richard Nixon during the late 1960s until Nixon's disgrace and fall from power during the Watergate scandal of the early 1970s. During his time as an operative for the Republican Party or GOP (Grand Old Party), Ailes had proposed the creation of a television channel or news service: the GOP on TV News, devoted to broadcasting messages and images from a socially conservative, business-friendly standpoint.

At the press conference that introduced Fox News, Roger Ailes proclaimed "We expect to do fine, balanced journalism."[2] But very soon the channel took an entirely different course, one very much in line with his political background. Under Ailes's forceful leadership, Fox News soon began to blur the boundaries between news, opinion, and blatant advocacy. Today the channel's comforting slogans – "Fair and Balanced" and "We Report, You Decide" – serve as mere cover for an approach to television info-tainment in which far right-wing political constructions of national and world events are the dominant theme.

A significant feature of Fox News that makes it attractive to today's television viewers is the use of engaging visual devices. On the screen at any given moment are two or more vivid, colorful banners – streams of moving text – that fill much of the bottom half of the picture as newscasters talk and videos of the day's events fill the other half of the frame. Typically, the content of the banners has little or nothing to do with the story or commentary depicted above. The idea is to barrage viewers with as much "information" and stimulation as they can possibly absorb. So effective is this approach in attracting TV audiences that other cable and broadcast companies have found it necessary to copy this busy, flashy, visual style. Especially during

[2]Roger Ailes at 6 min, 9 s in the video documentary, *Outfoxed: Rupert Murdock's War on Journalism*. 2004. Robert Greenwald, director and producer.

periods in which obviously important news is breaking – the expected and later actual outbreak of the US attack on Iraq in 2003, for example – the eye-catching text and pictorial images of political figures, aircraft, and fiery explosions are accompanied by portentous symphonic music that enhances feelings of danger, grandeur, and, perhaps, impending violence.

Fox News reports, talk shows and ongoing streams of vivid commentary typically present strong explicit or implicit distinctions between "us" and "them," "right" versus "wrong," "good" versus "evil." Thus, plans to build a community center for Muslims in lower Manhattan was consistently depicted as an ominous development that Fox named the "Ground Zero Mosque," supported by persons and groups that Fox implied were associated with the attacks on 9/11, in particular, believers in the Muslim religion. Commentators on the channel further often depicted Muslims as a threat to basic American values.

In its coverage of events of the day, Fox News eliminates the boundary between news and opinion, often featuring carefully prepared "talking points" issued by right wing organizations, ideological conservatives, and the Republican Party. This approach proved especially pungent during the build-up to the U.S. invasion and subsequent occupation of Iraq. Fox News enthusiastically echoed claims by the Bush administration that Saddam Hussein possessed weapons of mass destruction and was preparing to launch them at the United States. They also claimed that Iraq was the center of terrorism worldwide, and that it had links to those who planned and executed the attacks on the World Trade Center and Pentagon on September 11, 2001. The implication was that Iraq and its people deserved the death and destruction rained down upon them by the American military. Although none of the points in the Bush administration case against Iraq proved to be true, they offered a convenient way of identifying a scapegoat for the post-9/11 anger that many Americans felt. Never much concerned to probe the validity of the supposed "evidence," the on-screen personalities of Fox News took the lead in whipping up the nation's rage.

With the coming of the Barack Obama administration in 2009, much of Fox News coverage abruptly shifted to explicit attempts to undermine the credibility of the new president and his policies. Hence, Fox was the center of continuing rumors that Obama was not really an American, that he had not even been born on American soil, a bizarre belief (refuted by official birth certificates from the State of Hawaii) that came to be known as "birtherism." In contrast to the good, virtuous, hard-working people of the American heartland, President Obama was consistently depicted as a strange and threatening "other." Similarly, during the debates of 2009–2010 on health insurance reform legislation, Fox consistently described Obama's policies as a "Government Takeover of Health Care." Specific features of the bill, such as the provision of insurance for voluntary, end of life counseling, and help in the preparation of living wills, were described as a nefarious attempt to institute what Fox News described as "death panels."

Testimony of former Fox employees often show that bias, exaggeration, and fabrication of this kind are, in fact, products of deliberate, well-planned policies to supplant independent, professional, fact-seeking journalism with information tuned to a particular ideological position. As Jon Du Pre, the former Fox reporter in its West Coast Bureau, noted in an interview for the documentary *Outfoxed: Rupert Murdock's War on Journalism,* "We weren't necessarily, as it was told to us, a news gathering organization, so much as we were a proponent of a point of view."[3] Especially revealing in this regard are memos leaked by Fox employees that show an ongoing flow of instructions sent by Fox's top executives to their editors and news readers about acceptable and unacceptable ways to describe national and world events. For example, during the early stages of the Iraq war, Fox senior vice president John Moody sent out the following instructions about a piece of video footage: "Let's refer to U.S. marines we see in the foreground as 'sharpshooters', not as 'snipers' which carries a negative connotation."[4]

Pushing a right-wing belief that global warming does not exist and should not be addressed in government policy measures, Fox consistently derides scientific findings about climate change as uncertain and perhaps even a hoax. In December 2010, Fox News editor Bill Sammon sent an email to his colleagues saying: "… we should refrain from asserting that the planet has warmed (or cooled) in any given period without IMMEDIATELY pointing out that such theories are based upon data that critics have called into question. It is not our place as journalists to assert such notions as facts, especially as this debate intensifies."[5]

Neither easily verifiable facts in the day-to-day reporting, nor the well tested scientific consensus of consequential issues such as climate change, pose any barrier to the positions Fox reporters and pundits offer to their audience as "truth." When the weight of empirical evidence appears to discredit a position Fox News seeks to promote, its characteristic response is to raise questions about the motives and personal character of those offering the evidence, rather than to probe the factual basis of the claims in question. For example, when leaks of email messages written by climate scientists showed them deliberating over choices about which evidence to release in reports about global warming, Fox News name the communications "Climate-gate," a sure sign, in its view, that reports of climate change caused by human activity were a vast fraud perpetrated by the scientific community. Careful review of the actions of the scientists involved in the controversy later completely cleared them of any wrongdoing.[6]

A carefully orchestrated set of verbal gestures helps Fox News broadcasters maintain firm control of the network's message content. When a person being

---

[3] Jon Du Pre, at 7 min 49 s, *Outfoxed: Rupert Murdock's War on Journalism* (see complete reference in footnote 2).

[4] Memo from John Moody, 28 April 2004, quoted at 10 min, 15 s, *Outfoxed: Rupert Murdock's War on Journalism* (see complete reference in footnote 2).

[5] "Foxgate: Leaked email reveals Fox News boss Bill Sammon ordered staff to cast doubt on climate science" (capital letters in the original), in *Climate Science*, website edited by Joe Romm. http://thinkprogress.org/romm/2010/12/15/207201/leaked-email-fox-news-sammon-cast-doubt-on-climate-science/?mobile=nc. Accessed 15 January 2012.

[6] Nikki Fox, "Concerns over Climategate inquiry," *BBC*, 21 December 2011. http://www.bbc.co.uk/news/uk-england-norfolk-16294420. Accessed 15 January 2012.

interviewed seems to stray into topics or arguments the Fox host finds disagreeable, he or she will likely be interrupted, shouted down and silenced, sometimes with the rude phrase, "Shut up!" A somewhat more subtle way to turn the discussion toward controversies favored by Fox News occurs when an interviewer begins a question with the words, "Some people say that..." or "Many people in Washington are saying that...." In such cases, the actual source of the report – if, indeed, there is any at all – is never identified. The person asked to comment is, in effect, forced to take the assertion at face value. To ask, "Can you tell me who said it?" would seem a breach of confidentiality with the unnamed (quite possibly imaginary) source. Thus, whole televised discussions of political events are predicated on reports that may lack any real basis at all.

The range of acceptable opinions on Fox News typically includes a mix of socially conservative, "free market," libertarian, narrowly traditionalist, fundamentalist and evangelical Christian, anti-black, anti-gay, anti-immigrant, American nationalist, militarist, and, most prominently, corporatist worldviews. Any ideas that depart from this range of positions are usually depicted as strange, wrong-headed, or simply contemptible. As a way to implant its ideas even more deeply, Fox routinely employs subtle, even subliminal ways of coding its messages. On occasions in which a politician from the much preferred Republican Party has been caught in a financial or sex scandal, the text on the screen will place a "D," for Democratic Party, after the person's name, a "mistake" that happens often enough to be recognized as something more than inadvertent. By the same token, Fox is known to alter photographs of political figures it wishes to discredit, by adding dark circles under their eyes, for example.[7]

The extent to which images can be manipulated to match the company's line extends to the use of graphs that misrepresent factual data. During a period in late 2011 in which unemployment fell from 9 to 8.6 % indicating that the Obama administration's economic policies were having trouble producing favorable results, the Fox News display of the data's month-to-month trend line showed the 8.6 % number at a higher point on the graph than figures of 8.8 and 8.9 % earlier in the year.[8] The impression was, therefore, that there had been no improvement in the numbers at all. In an even more astonishing case, Fox depicted poll results for a number of Republican candidates vying for the party's 2012 presidential nomination. Leading in the poll at the time were Newt Gingrich with 29.3 % and Mitt Romney with 17.2 % with the data and photos shown side by side. Alas, the photo over the name "Romney" was actually that of Barack Obama, an effect that suggested Gingrich was actually leading Obama in the polls. Again, while blunders of this kind are bound to happen occasionally in any newspaper or television report, they happen so frequently on Fox News as to support the conclusion (along with other evidence of bias and misrepresentation) that they are a product of design rather than sloppy workmanship.

---

[7] A description of instances of this kind are presented in "Fox News Channel controversies," Wikipedia. http://en.wikipedia.org/wiki/Fox_News_Channel_controversies. Accessed 15 January 2012.

[8] "Today in Dishonest Fox News Charts," *Media Matters*, website, 12 December 2011. http://mediamatters.org/blog/201112120005. Accessed 13 December 2011.

There seem to be no rhetorical moves too excessive or comparisons so extreme that they cannot be employed on Fox News. The very fact that the discussions are beyond reason and colorfully presented is what gives them such strong audience appeal. For example, on August 26, 2003, about 6 months into the War in Iraq, Fox News's most prominent veteran anchor, Brit Hume sought to diminish the impression that the cost of the war was too high as measured by the number of U.S. soldiers killed in battle. "Two hundred seventy-seven U.S. soldiers have now died in Iraq," he noted, "which means that statistically speaking U.S. soldiers have less of a chance of dying from all causes in Iraq than citizens have of being murdered in California, which is roughly the same geographical size."[9] The utterly ludicrous comparison in Hume's statement conveys the underlying strategy in Fox News presentation of national and world events – that brashly, colorfully expressed opinions far outweigh any verifiable facts.

To a great extent, the success of Fox News stems from the ways it appropriates and modifies the legitimate, tried and true traditions of journalism from earlier times. At first glance the reassuring features of news reporting and thoughtful editorial comment are fully present. Yes, there are desks with men and women looking into the camera with stern expressions, talking about what's happening in the "news." Yes, there are segments from correspondents in distant locations seeming to provide greater detail and depth on breaking stories. And, yes, there are groups of "experts" who offer their best knowledge about the broader significance of issues covered that day. But seen from a more exacting, critical standpoint, much of what appears on the screen is pure spectacle driven by the need to propagate a preconceived set of ideological conclusions.

The most trenchant observations about Fox News's mode of operation come from those who were themselves willing participants in "conservative" news and publishing during the years in which Rupert Murdock organized his television operations in the United States. One of them, David Brock, a dirt-digging scandal-monger during the 1990s known for his stories about the supposed misdeeds of Bill and Hilary Clinton, later renounced his unscrupulous work and its millionaire sponsors and went on to found Media Matters, a watchdog organization that keeps close tabs on Fox and other sources of right wing "news." Interviewed in "Outfoxed," Brock derides Roger Ailes for deliberately trashing the methods and standards of the profession of journalism.

> He doesn't believe in objectivity. He has contempt for journalism, I think. He wanted all news to be a matter of opinion because opinion can't be proven false. I think that is very dangerous because if people don't have a set of facts that they can agree upon, I think it's difficult to reach a consensus on what's correct public policy.[10]

Despite these features, or perhaps because of them, the specific programs on Fox regularly top the ratings for cable news television. Among all cable channels of

[9]Brit Hume quoted in "Al Franken, God Spoke: Divining Intervention in Politics," Louis Proyect, WBAI.org. http://wbai.org/index.php?option=com_content&task=view&id=9080&Itemid=2. Accessed 15 January 2012.

[10]David Brock, at 16 min, 34 s. *Outfoxed: Rupert Murdock's War on Journalism* (see complete reference in footnote 2).

every variety in America, Fox ranks fourth overall in the size of its audience. In the "prime time" evening hours it is consistently first among news providers. Its main challengers, CNN and MSNBC, lag far behind in audience share. In 2010 some 41 million viewers watched at least 50 min of Fox News each month (Holcomb et al. 2011). Within the larger picture, the older, established networks of broadcast television – NBC, CBS and ABC, still command a larger viewing audience. But as regards the power of television to influence the agenda for political discussion and content of public policy, Fox is now clearly the leading force.

The phenomenal success of Fox News is reflected in both its ratings and profits from the sale of advertising. As a result, other television news organizations, especially CNN, have begun copying the visual, sonic and rhetorical style of the channel along with its topical and ideological content. Thus, in one notorious episode, Fox decided that digital images of a particular Democratic Party Congressman's sexual organs foolishly sent over Twitter (to a woman not his wife) should become the leading news lead story for days on end. That prompted the other networks to emphasize the seedy report as well, rather than risk losing their share of the viewing audience. Within media stampedes of this kind, more and more of American current-events programing dwells (or drools?) upon celebrities, scandals and fleeting fashion trends, distracting attention from what are arguably more crucial matters. As social critic Chris Hedges observes,

> Those captivated by the cult of celebrity do not examine voting records or compare verbal claims with written and published facts and reports. The reality of their world is whatever the latest cable news show, political leader, advertiser, or loan officer says is reality. The illiterate, the semiliterate, and those who live as though they are illiterate are effectively cut off from the past. They live in an eternal present (Hedges 2009: 47).

As it has emerged in recent years, one purpose of the Fox propaganda machine is to influence both the elections and operations of the nation's government directly, something that goes beyond merely shaping the tone and content of public debate. This is accomplished by serving as a hot-house where candidates for the presidency and other high offices are supported, nurtured and given a highly visible media position from which they can launch the next steps in their political careers. As paid consultants, pundits and regular participants on Fox News programs hone their rhetorical skills and position themselves for leading roles in government itself. At this writing, among actual or potential candidates of this kind are Sarah Palin, Newt Gingrich, and Mike Huckabee. So powerful is Fox News' role as kingmaker in the Republican Party that at the beginning of the second decade of the twenty-first century it appeared to many observers that the conventional relationship between politics in the real world and commentaries within the realm of electronic media had, to great extent, been reversed. Pushing a far right wing agenda – anti-abortion, anti-union, anti-immigrant, anti-public education, anti-Social Security, anti-Medicare, anti-climate change science, anti-public spending on and all social programs, etc. – Fox News had become the wellspring and center of decisions for the Republican Party itself, ready to realize its power directly within Washington, D.C. as well as many state governments.

It may seem implausible that Rupert Murdock's television channel could have achieved such extraordinary power in barely 15 years of operation. But the mathematics of the American elections help make this result entirely feasible. As reflected in the victories of George W. Bush in 2000 and 2004, Fox News and the conservative base of the Republican Party appear able to command a solid, enthusiastic "base" of support that amounts to 25–30 % of the U.S. populace that is likely to vote. That leaves only a margin of 20 % of the rest of voters who have to be swung to the support of a conservative Republican candidate if that person is to win an electoral majority. In a nation in which about half of the eligible population does not bother to vote at all, it is well within reach for a relatively small faction of zealous partisans, mobilized by a dedicated, state-of-the-art propaganda factory, to succeed in choosing who will hold the highest office in the land.

The growing political muscle of the Fox News Channel was starkly displayed in the congressional elections of 2010. A pseudo-populist movement, the Tea Party, largely promoted and supported by Fox and trumpeted by the channel's anchor personalities, was able to galvanize discontent with the presidency of Barack Obama and spiraling levels of government debt, enough so that the election moved the control of the House of Representatives to the Republican Party. For the remainder of Obama's term as president, opposition from Tea Party congress members prevented the passage of any significant legislation to address the country's problems. The major objective Tea Party Republicans sought to achieve was "for President Obama to be a one-term President."[11]

In a speech delivered in Springfield, Missouri 3 years before the outbreak of the American Civil War, Abraham Lincoln (1858) proclaimed, "I believe this government cannot endure, permanently half slave and half free." Surveying the kinds of poisonous discourse and political wreckage Fox News Channel has perpetrated, a prominent blogger who writes under the pseudonym "driftglass," recently rephrased Lincoln's jeremiad: "The nation can no longer survive half-Fox and half-free."[12]

3 Democracy and Propaganda at an Impasse

Ellul's central contention – that people in modern society have an intense craving for propaganda in their every waking moment – seems fully borne out in the development of Fox News. Despite its widely noted record of inaccuracies, distortions, and outright lies, surveys show that about half of the U.S. television

[11] "Sen. McConnell: Making Obama a One-Term President is my Single Most Important Political Goal," *Mediaite*, 10 July 2011. http://www.mediaite.com/tv/sen-mcconnell-making-obama-a-one-term-president-is-my-single-most-important-political-goal/. Accessed 15 January 2012.

[12] "Only Nixon can go to Nixonland," driftglass, blog, 26 July 2011. http://driftglass.blogspot.com/2011/07/only-nixon-can-go-to-nixonland.html. Accessed 15 January 2012.

viewing audience finds the channel to be the "most trusted" among the choices available.[13] Perhaps this means that the picture of the world Fox News projects each day can be trusted to reflect deeply held beliefs that many people cling to in troubled times. Fox News helps its audience imagine a world in which a well policed electric fence prevents Mexican immigrants from crossing the southern border; a world in which the threat of climate change is revealed to be mere rumor; a world in which gay couple's plans of getting married are outlawed; a world in which Muslims must realize that they are not welcome; a world in which meddling "liberals" and their dreamy programs of social reform will be stopped in their tracks and "Big Government" eliminated from any influence in ordinary people's lives. In short, Fox can be trusted to reflect back to its viewers their most deeply held desires, fears, resentments, myths, and hatreds, presented around the clock as superficial, enjoyable info-tainment and "news you can use."

The account Ellul offers to explain the ultimate appeal of "news" of this sort also seems to fit the situation in which many Fox News viewers find themselves. Economic and social transformations seen everywhere on the planet obviously threaten their traditional ways of living. Pressures of globalization, job loss, personal debt, shattered families, fragmented local communities, and the arrival of people of different races, cultural backgrounds, and sexual orientations are profoundly unsettling for those taught to revere the stable values of 1950s suburban America. Faced with growing upheavals of this kind, Fox News presents the hope of restoring social and political patterns now in decline. It also indulges widespread fantasies of striking back at groups of alleged wrongdoers conveniently blamed for the maladies that have stricken the US during the past three decades.

Unfortunately, as regards the contribution Fox News makes to the levels of information and understanding about current events its audience commands, the evidence is ominous. Social scientific polls that measure how much various groups of regular television viewers know about important events in the news indicate that Fox viewers command far less accurate knowledge than those who watch other television channels.[14] In fact, some surveys show that Fox viewers actually have far less information about important national and international developments than people who watch no television at all.[15] The old adage of media critic Danny Schecter seems to apply: "The more you watch, the less you know."

Since the early days of the George W. Bush administration and the expanding influence of Fox News, several varieties of resistance to the onslaught of right-wing propaganda have taken shape. As a way to test and counter the daily barrage of claims that issue from Rupert Murdock's shop, several organizations have begun

[13]"Poll: Fox most trusted name in news," Andy Barr, Politico, website, 27 January 2011. http://www.politico.com/news/stories/0110/32039.html. Accessed 1 December 2012.

[14]"Voters Say Election Full of Misleading and False Information". World Public Opinion. Org, 9 December 2010. http://www.worldpublicopinion.org/pipa/articles/brunitedstatescanadara/671.php?nid=&id=&pnt=671&lb=. Accessed 30 May 2011.

[15]"Some News Leaves People Knowing Less". Farleigh Dickinson University Public Mind Poll, 21 November 2011. http://publicmind.fdu.edu/2011/knowless/. Accessed 1 December 2011.

conducting intense, detailed fact-checking and analysis of the various reports the company presents as unquestionably true, "fair and balanced." The aforementioned Media Matters maintains a 24-hour-a-day watch over the statements, images and overall themes that appear on the channel, asking how well they stand up to critical scrutiny. Crooks and Liars, an Internet web site, pays particular attention to Fox video segments, flagging the amusing excesses and falsehoods they sometimes contain. In addition, several newspapers, including the *Washington Post*, have instituted fact-checking services to verify news stories and statements of public figures, not just on Fox airwaves but in the print and electronic sources more generally. A sad recognition that inspires initiatives of this kind is the abrupt decline in standards in what passes as journalism today as compared to earlier times. While reporters of previous generations were expected to locate at least two reliable sources before publishing a claim as factual, today's "journalists" often rush to the TV cameras with rumors from a single, highly suspect informant. Among professionals in the field there is widespread recognition that Fox News has led the way in degrading the norms that guide news reporting and commentary.

Another promising strategy for restoring credibility and integrity in television journalism has been to organize alternative cable and satellite channels that carry a more "liberal" and factually reliable slant in reporting and interpreting the news. While such efforts are often strongly influenced by the priorities and management of profit-seeking media corporations, there are some indications that more credible news practices and somewhat more diverse sources of opinion will be made available to TV audiences. Among the evening cable programs such as MSNBC, are ones hosted by a young woman with a PhD in political science, an African American activist, an experienced journalist who was formerly an aide to a Democratic Party leader in the House of Representatives, and a talk show host sympathetic to the interests of labor unions and working people – all of them much different in character and much more careful in their presentation of political issues than the mainly white, corporatist, "conservative" talkers on Fox News. Other channels featuring strong investigative journalism and commentaries with a liberal or radical slant include Free Speech TV, Link TV and Current.

For both models of media reform – improved fact checking and the development of superior television channels – a strong concern is to provide information and discussion more consistent with the needs of a vibrant twenty-first century democracy. By eliminating the varieties of "faux news" and "toxic talk" commonly found on Fox News and replacing them with more substantive sources of news and opinion, organizers of the new outlets hope that a better informed, critically minded public can be nurtured and that, eventually, a new wave of intelligent, caring public officials will achieve positions of power.

From Ellul's standpoint, reforms of this kind express a largely misguided hope that somehow today's democracies can evade or perhaps repair the corrosive influence of propaganda in our political institutions. Thus, he advises that scholars and citizens use extreme caution as they yearn for remedies of this kind. "Historically," he writes,

> from the moment a democratic regime establishes itself, propaganda establishes itself alongside it under various forms. This is inevitable, as democracy depends on public opinion

> and competition between political parties. In order to come to power, parties must make propaganda to gain voters (Ellul 1965: 232).

The very point of propaganda in all its form, he argues, is to be effective in inculcating beliefs, ideologies, and frameworks of understanding that tend to neutralize the kinds of open inquiry, discussion, and debate that genuine democracy involves. The hope that one can simply try to identify the lies and correct the distortions is bound to be forlorn because it misunderstands the pervasive presence of propaganda in every attempt to move a mass populace. Aspirations of that kind also tend to delude us into thinking that we are somehow special, virtuous, and truthful, and, therefore, can be inoculated from the plague.

For those who care about public life, the paradox Ellul spells out is a vexing one. In large modern societies democracy depends on the skillful use of propaganda to mobilize the populace for political ends. But in order to be effective in achieving the political goals its users seek, propaganda must inevitably seek to neutralize democracy's distinctive strengths, especially the original thoughts, civic deliberations and decision-making initiatives of everyday citizens. As he outlines the fundamental tension, "some of democracy's fundamental aspects paralyze the conduct of propaganda. There is, therefore, no 'democratic' propaganda. Propaganda made by the democracies is ineffective, paralyzed, mediocre" (Ellul 1965: 241). By the same token, he argues, "With the help of propaganda one can do almost anything, but certainly not create the behavior of a free man or, to a lesser degree, a democratic man" (Ellul 1965: 256).

Comments of this kind pose the question: Who does Ellul think "a democratic man" actually is? His answer, although not lengthy, is entirely clear, clarified by the threat propaganda presents. "A man who lives in a democratic society and who is subjected to propaganda is being drained of the democratic content itself – of the style of democratic life, understanding of others, respect for minorities, re-examination of his own opinions, absence of dogmatism" (Ellul 1965: 256).

To my way of thinking, Ellul points to the validity of a person's direct experience, of immediate rather than mediated contact with other people, of inquiries and discussions that bring to focus the most basic questions about our shared existence and common commitments. Anything that interferes in experiences and activities of that kind is bound to weaken and undermine the lives of authentically democratic citizens. Ellul insists that perhaps the most debilitating interference for the realization of democracy in our time is propaganda in all its forms.

> The means employed to spread democratic ideas make the citizen, psychologically, a totalitarian man. The only difference between him and a Nazi is that he is a 'totalitarian man with democratic convictions,' but those convictions do not change his behavior in the least. Such contradiction is in no way felt by the individual for whom democracy has become a myth and a set of democratic imperatives, merely stimuli that activate conditioned reflexes (Ellul 1965: 256).

Ellul's argument evokes the most notorious historical manifestations of propaganda in totalitarian regimes to remind us that what seem to be fairly innocuous, everyday varieties of information management, entertainment and public opinion shaping are, upon closer inspection, incompatible with genuine, authentic, directly democratic modes of democratic life. As I have suggested, Rupert Murdock's Fox News is a living example of the menace Ellul describes – engaging, entertaining, and full of the messages people want to hear. The book alerts us to the distinct possibility that many of the proposed correctives to the influence of Fox in the public sphere – better fact checking and the creation of more responsible yet lively sources of news programming – will simply compound the basic malady.

Faced with conditions of this kind, what is to be done? As is characteristic of his writing, Ellul pulls no punches. His advice is to rise up, sound the alarm and alert one's fellow citizens of the menace that confronts them.

> The only serious attitude – serious because the danger of man's destruction is serious, serious because no other attitude is truly responsible and serious – is to show them the extreme effectiveness of the weapon used against them, to rouse them to defend themselves by making them aware of their frailty and the vulnerability, instead of soothing them with the worst illusion, that of a security that neither man's nature nor the techniques of propaganda permit him to possess (Ellul 1965: 257).

For people like me, those engaged in political activity and who think of themselves as thoughtful, democratic citizens, Ellul offers a challenge of a most unsettling kind. If any attempt to spread ideas widely within the public realm veers toward totalitarianism, what can one say about one's own writings, talks, posters, petition drives, marches, and the like? Does the very attempt to attract substantial numbers of people to one's point of view through persuasive rhetoric or writing make one a propagandist? Can one appear on television to offer information or make an argument without diving into the sewer of media chicanery?

The simple point of wisdom at the conclusion of *Propaganda* seems to be that all of us must rely on our direct experience, on our own inquiries, on our own sense of the world as we make judgments about important social, economic and political issues. Citizens must diligently avoid pre-packaged, pre-cooked versions of reality offered by media technicians, corporate managers and anyone brandishing a firmly fixed ideology. Given the barrage of misinformation that bombards the populations of media saturated societies around the globe, Ellul's counsel offers a small but potent ray of hope for the future of democracy.

References

Ellul, Jacques. 1965. *Propaganda: The formation of men's attitudes*. Trans. Konrad Kellen and Jean Lerner. New York: Knopf.

Hedges, Chris. 2009. *Empire of illusion: The end of literacy and the triumph of spectacle*. New York: Nation Book.

Holcolm, Jesse, Amy Mitchell, and Tom Rosenstiel. 2011. Cable: By the numbers, *Pew research center's project for excellence in journalism: An annual report on American journalism 2011*. http://stateofthemedia.org/2011/cable-essay/data-page-2/. Accessed 1 Dec 2011.

Kornhauser, William. 1959. *The politics of mass society*. New York: Free Press.

Lincoln, Abraham. 1858. *House divided speech*. Speech to the Republican State Convention, Springfield, Illinois, 16 June 1858. http://showcase.netins.net/web/creative/lincoln/speeches/house.htm. Accessed 15 Jan 2012.

Chapter 9
An Unseasonable Thinker: How Ellul Engages Cybercultural Criticism

Andoni Alonso

> *I do not know what meaning classical scholarship may have for our time except in its being 'unseasonable,' that is, contrary to our time, and yet with an influence on it for the benefit, it may be hoped, of a future time.*
>
> Frederich Nietzsche
> *On the Use and Abuse of History for Life*

> *The real issue is that humans are no longer in charge. We need to dismantle the machines themselves. This can be done in a very peaceful manner. Hack into their system, publish their crimes through Wikileaks-type initiatives and then delete their real-time trading killing networks for good.*
>
> Geert Lovink and Franco Berardi
> *A Call To the Army of Love and To the Army of Software*

1 Ellul in Brief

Jacques Ellul is a scholar difficult to classify. His more than 40 books and hundreds of articles have contributed to theology, sociology, history, and economics. Today in the era of the Internet, global communications, and the dominance of technology, Ellul is often dismissed as a techno-catastrophist or misleading heretic. Also labeled a Christian "neo-luddite," Ellul did indeed produce an analysis of contemporary technology as potentially leading to catastrophe – and few people are pleased by

A. Alonso (✉)
Facultad de Filosofía y Letras, Universidad Complutense de Madrid
Ciudad Universitaria, Madrid, Spain
e-mail: andoni.alonso@pdi.ucm.es

H.M. Jerónimo et al. (eds.), *Jacques Ellul and the Technological Society in the 21st Century*, Philosophy of Engineering and Technology 13,
DOI 10.1007/978-94-007-6658-7_9,

such criticism, especially when the economy appears to grow without limits and there are more and more goods for consumption. In a "low cost" culture technological criticism is not an easy sell. Until recently we lived in a world of the *Apocalypse Postponed* (Eco 1994).

Technological enthusiasm obscures the relevance of Ellul's thought. Ellul has contributed, along with others such as Lewis Mumford and Ivan Illich, to a current of criticism that has been a strong if hidden influence on others such as Paul Virilio and Jean Baudrillard. "Certainly there are echoes of Ellul's technique, of the negative effects" on "the organization of social life and space from the First World War to the present day" (Armitage 2011: 5); and despite his Catholic Christianity, Virilio himself sees his thinking as closer to Ellul than to, for example, Gabriel Marcel (Armitage 2000). A "postmodern" appreciation of Ellul's thinking deserves consideration.[1]

Ellul was neither luddite nor technophobe. He simply emphasized the ambivalent nature of the technological phenomenon. Given that the recent financial crisis may be undermining an apocalyptic postponing mentality, perhaps Ellul's criticism can appear in a more favorable light. The European way of life, from human rights to the welfare state and social protections, seem to be at risk. Certainly the myth of progress is being seriously questioned and new ideas are needed to reinvent ourselves. According to Naomi Klein (2008) we live in the midst of a capitalism that is characterized by its commitment to a "shock doctrine." That is, it promulgates a series of crisis by which excessive speculative profits can be accumulated. Such disaster capitalism is spreading, with its shock characteristics facilitated by technologies that are out of control.[2] Yet even after Chernobyl and Fukushima, criticisms of technology are seldom acknowledged and may still be ignored.

Nevertheless, for the first time we hear that the lives of future generations could be worse than previous ones. Global financial meltdown, high unemployment rates, and environmental degradation indicate something about how the economy and the technology that empowers our financial milieu are counter-productive for our lives – or that they are mostly benefiting only a few, and that even this elite is shrinking. According to the Occupy Wall Street movement, 1 % of the US population receives three quarters of the total income increases of the country

[1]Thanks to STS groups in Spain and through the involvement of scholars such as Carl Mitcham and Langdon Winner, Ellul has become part of the Spanish repertoire on philosophy of technology. Indeed, Ellul has been extensively translated into Spanish. Ten books (including *The Technological System*) and some essays of have been translated as recently as 2011. See Florensa Giménez (2010), Sanchís Serra (2009, 2011).

[2]Computer scientists, mathematicians, engineers, even linguists, are now part of the financial casino. And they along with technologists are responsible for the present crisis. As Leinweber says: "The Hall of Shame for those guilty of incompetent engineering features collapsing bridges, flaming dirigibles, exploding spacecraft, and melting reactors. We can add a new wing for overly complex [financial] derivatives, modelled in exquisite detail by myopic nerds with Ph.D.'s who got lost in the ever more complex simulations but ignored the basic principles, and their lavishly paid bosses who ignored the warnings from the best of them so they could be even more lavishly paid" (Leinweber 2009).

(Luhby 2011). Ellul considered himself as a resistor and frequently repeated the need to stand against the general state of affairs. For Ellul, it was important to say that the actual combination of politics, economy, and technology are creating more harm than benefit.

Another relevant issue is that for Ellul an intellectual position is not simply a "detached point of view" for describing how things are. Ellul understood intellectual activity as a way to warn, denounce, and criticize. These elements are often absent from scholarly papers today. What are the consequences? The answer is revealed in one of the most upsetting documentaries on the present economic crisis, the Oscar-winning *Inside Job* (2010), in interviews with economists from Harvard, Berkeley, and other premier universities. These economist scholars demonstrate how ethics, social compromise, and honesty have disappeared or been ignored. Economists in academia have supported an intense propaganda favoring financial networked capitalism and deregulated free market policies. Are such attitudes not also characteristic of social sciences and humanistic discourses on technology? Does the scholarly study of technology not too often support the current, deleterious state of affairs? Have we scholars simply adopted positions that would allow us to get funds and recognition? Detachment was certainly not the case in Ellul's thinking. Radical criticism of technology may be one reason why Ellul – along with Illich, Mitcham, Winner, and others – should be considered "unseasonable thinkers" – free spirits who go against the *Zeitgeist*, trying to identify ways forward to a better life in the contemporary world.

Unseasonable thinkers are not detached from questions and problems in contemporary society. Just the opposite: They criticize general assumptions such as progress, growth, innovation, and the like. They expose and confront general myths. Practical answers for what to do or counter-ideologies may not be available. But while the absence of answers can be dissatisfying, problems can still be sharply presented. What is interesting about Ellul in this regard is how others who may not have read his work nevertheless echo his ideas. Ellul was touching central concerns in contemporary society. A further review of some of these ideas and how they engage with cyberculture can enhance appreciation of Ellul as well as illuminate critical discourse regarding information and computer technology. The focus will be on three themes – the sacred, speed, and work – all of which play prominent roles in cyberculture discourse.

2 Technology and the Sacred

According to Ellul, when people assert that technology "is an instrument of freedom, or the means to ascent to historical destiny, or the execution of a divine vocation" the result is the "glorifying and sanctifying of Technique." Technology ceases to be

> an ensemble of material elements, [and becomes instead] that which gives meaning and value to life, allowing man not only to live but to live well. Technique is intangible and

> unattackable precisely because everything is subject and sub-ordinate to it. Man unconsciously invests with a holy prestige that against which he is unable to prevail (Ellul 1962: 410).

Decades before others, Ellul identified technology as the new sacred, a theory that has since been reiterated by others (Bloom 1997; Wertheimer 1999; Noble 1999; Alonso and Arzoz 2003). Computer development and computer mediated culture – along with a host of other leading-edge technologies such as biotechnology and nanotechnology – have been repeatedly valorized by their proponents with a rhetoric that is religious if not superstitious. In the last 20 years of the Internet, bizarre appeals have been made to digital networks as exhibiting omnipresence, the power to save humans from pending disasters, and to raise them to a transcendent ontological status.

Anticipating such transcendence, the priest-scientist Teilhard de Chardin, in affirmative speculations developed during the same period as Ellul's social criticism, proposed that technological progress would bring humanity to another level of spirituality (Teilhard de Chardin 1959) – a vision echoed in the "omega point" allegedly discovered by Frank Tipler (1997) and the "spiritual machines" of Ray Kurzweil (2000). In these visions the body, nerves, psycho-social conditions, and other features of human presence in the material world appear as obstacles to achieving a promised transformation. Sacrificing the body to the mind is the paradoxical requirement for receiving the blessings of a technological utopia and the eschatological moment. Although referring originally to how God will bring about the end of the world and what he will preserve in a post-end state, eschatology has come through science fiction to reference a world-end immanent in technological history as it moves toward the singularity in which humans will be overcome and saved by their machines – see, e.g., Stross (2003), Stephenson (2000), and Egan (1994).

To some extent modern science involved this idea from the beginning. According to Francis Bacon, science and its associated technologies are liberating forces able to defeat nature and place humanity on a new footing in the world. This ideology translates into the present time when scientists (Tipler 1997), engineers (Kurzweil 2000), intellectuals (Lèvy 1994), and artists (Stelarc 2011) advocate for a cyber-spirituality that sees humanity returning to Paradise by means of miraculous technoscientific achievements. These ostensibly new visions incorporate from the hermetic tradition such myths as those of the new body and immortality and, above all, the creation of a virtual and artificial god. This ersatz religion, which is closely associated with information technology, has been termed "digitalism" (Alonso and Arzoz 2003) and displays affinities with ancient Gnosticism. Indeed, digitalism is a sort of "techno-hermetism" that incorporates a mixture of crude rationalism, esoteric religious traditionalism, and science fiction in order to map out techno-epistemic pathways to revelation and transcendence. Together with the mystifications of globalization, digitalism further promotes cultural uniformity and a pseudo-religious aristocracy. Cyber-intellectual missionaries, novelists, science fiction filmmakers, Internet cyber-sects, enthusiastic scientific journalists, and a growing number of contributing scholars celebrate transformations of human experience in terms of limitless scientific abuse. The movie *Gattaca* (1997) skillfully explores this abuse related to biotechnology. Extropians, post- and trans-humanists, Raelism and similar cults place faith in technology as a sacred means to achieve a transcendent end. Ellul's analysis of propaganda (Ellul 1973) could fruitfully be applied to much cyberculture hype.

Ellul's postulate that technology has become a kind of sacred is thus alive and well in cyberspace. For many cyber-intellectuals, virtual realities and online social environments are magical realms where intelligence alone is present and it is possible to speak about a collective entity called the mind-hive (Lèvy 1994). Cyberspace replaces the corporeal body with networks and processors. Cyber-believers imagine computers as the ultimate destiny of humans. Marvin Minsky, a father of artificial intelligence, once described the human brain as a "meat computer," reversing the metaphor of the "mechanical brain." In 1970 he even made the now-dated prediction that

> In three to eight years we will have a machine with the general intelligence of an average human being. I mean a machine that will be able to read Shakespeare, grease a car, play office politics, tell a joke, have a fight. At that point, the machine will begin to educate itself with fantastic speed. In a few months it will be at genius level, and a few months after that its powers will be incalculable (quoted in Stork 1996: 19).

Yet as Ellul would remind us, the people using computers are made of bones and flesh and blood. Cognitive workers are more than intellect: Cognitarians (to use a term from Berardi 2005) are not just nerves but somatic creatures with anatomies and physiologies that are repeatedly stressed by the continuous non-act of staring at screens. And collective intelligence cannot solve the problems of the social existence of those bodies that produce such intelligence.

3 Technological Speed

Discourse about the speed of technological change divides into concerns about how to accelerate it (we need more innovation) or how to limit it (everything is going too fast). For Ellul, limitation is the more crucial issue because we are physical, biological, and psychological creatures for whom speed is often disruptive. Many people readily agree that experiences of speed can distort and become sources of suffering. Individuals have increasingly been forced to adapt to an accelerated pace of life and social expectations; time seems to be one of the scarcest resources in post-modern societies.

Analyzing speed has been one of the classic paths to understanding and criticizing technological development. Intuitively, most people consider the technological pace as something troubling or even as something wrong. Illich's conviviality (Illich 1974) and Baudrillard's escape velocity (Baudrillard 1988) are different approaches to this problem. Past a certain point, speed is iatrogenic, as Illich argued. Virilio is perhaps the most well-known author here, beginning with his concept of *dromology* about the relationship of speed to politics (Virilio 1977). As Ellul shows, the effort required for human adaptation to breakneck change is an important issue related to others such as programmed obsolescence, increased consumption, sudden environmental changes, etc. Discussions about speed are especially difficult because they demand a suspension of the present trajectory and a rethinking of where we are going. To stop for a moment and reflect is the antithesis of speed.

It is difficult to deny that speed has become a crucial commodity. One of the most expensive office places to rent is close to the Wall Street stock market in New York City. Wall Street is the core for global financial activity. What is surprising, though, is that the price is not for individuals but for machines. In fact new technologies allow people to invest and operate in the Stock Market from almost any place in the world. But the closer the machine is connected to Wall Street the quicker the response time. According to technicians, time saved is about a millisecond less than if you have a machine connected in Madrid or Lisbon. This speed means advantage and money. But it also means that no human can compete with the speed of machines, and often no one can predict or foresee what is going to happen. This exemplifies, among other facts, the role of speed in our technological society. What counts is not the intelligence of a human operator dealing, reflecting, and making decisions. What counts is the power of a machine reacting as quickly as possible according to built-in heuristics. When something goes wrong the result can be as expensive as billions of dollars lost in 20 min, as happened on Wall Street on May 6, 2010 (Gilles 2010). Financial regulators are now forced to deal with what is called "flash trading".[3] The time able to operate in flash trading ranges from 300 to 500 ms. No human operators can control these operations; machines take command of such financial activities.

In cyber-organized society, computer technology becomes pervasive and invades all human life, as Franco Berardi (2009) points out. In this system, the question of speed becomes an ever bigger problem, as Ellul forecasted. According to Berardi:

> Semio-capital puts neuro-psychic energies to work, submitting them to mechanistic speed, compelling cognitive activity to follow the rhythm of networked productivity. As a result, the emotional sphere linked with cognition is stressed to its limit. Cyberspace overloads cybertime, because cyberspace is an unbounded sphere whose speed can accelerate without limits, while cybertime (the organic time of attention, memory, imagination) cannot be sped up beyond a certain point – or it cracks. And it is actually cracking, collapsing under the stress of hyper-productivity. An epidemic of panic and depression is now spreading throughout the circuits of the social brain. The current crisis in the global economy has much to do with this nervous breakdown. Marx spoke of overproduction, meaning the excess of available goods that could not be absorbed by the social market. But today it is the social brain that is assaulted by an overwhelming supply of attention-demanding goods. The social factory has become the factory of unhappiness: the assembly line of networked production is directly exploiting the emotional energy of the cognitive class (Berardi 2009: 276).

[3] *Flash trading* is a "controversial computerized trading practice offered by some stock exchanges. Flash trading uses highly sophisticated high-speed computer technology to allow traders to view orders from other market participants fractions of a second before others in the marketplace. This gives flash traders the advantage of being able to gauge supply and demand and recognize movements in market sentiment before other traders." http://www.investopedia.com/terms/f/flash-trading.asp#ixzz1fwpbTuky. Accessed 10 December 2011.

Concepts that articulate Berardi's idea are automation, speed, acceleration and general intellect. High technology workers have become the new "proletariat" but in a more profound sense than industrial labor, because their alienation is more pronounced. Workers in the assembly line were the first proletariat in history and now they have moved into underdeveloped countries where collective bargaining organizations have little power because of the political situations. In Western society labor is composed mainly of high-tech intellectual laborers – what George Orwell called "brain-workers"[4] – connected into networks in different categories such as techno-workers, networkers, and knowledge workers. The result is a cognitive capitalism that, in the last analysis, is the supreme way to alienate workers from their productivity.

Hackers and science fiction writers often distinguish among hardware, software, and wetware. Hardware are the machines, software the programs of machine operation. Wetware refers to the human programmer, administrator, or IT manager who operates with hardware and software. The logic of production is then to extract knowledge from humans and to do so at a low cost. There is a constellation of names for this cognitive capitalism: weightless economy, net-economy, new economy, information economy, digital economy, e-economy, knowledge value revolution or knowledge-based economy. This new-economy discourse coincides with notions that the Internet and telecommunications have become the most important realms for leisure and entertainment. Culture becomes a techno-logical realm, as does leisure (Aiestaran 2010).

This parallelism points toward how work and leisure are more and more in the same place so as to become less distinguishable. Step by step, the distinction between what is work time and free time begins to disappear. Is it possible to think that the mere use of a computer, no matter what the purpose, becomes a subtle way of working? Everybody knows that using a search engine such as Google becomes an added value for the software company. Users refine and add value to Google's system each time they use it, increasing the economic value to the company. The same happens with social networks such as Facebook or Linkedin; the value for these companies is a result of the time users spend with them. Those e-social networks are powerful devices to extract and sell information provided by people. What these social e-networks do is to transform normal conversations, and the banality of everyday life, into new business opportunities. Advertising has found another way to intrude into life through the use of "customized" ads using the knowledge obtained from users via ubiquitous data mining.

Even within what we can consider "traditional jobs" – jobs that have a regular wage and time-at-work – we see features that redefine them in cognitive capitalist terms: permanent and never ending training (also labelled as continuing education),

[4]"We pigs are brainworkers. The whole management and organization of this farm depend on us. Day and night we are watching over your welfare. It is for your sake that we drink that milk and eat those apples. Do you know what would happen if we pigs failed in our duty? Jones would come back! Yes, Jones would come back! 'Surely, comrades,' cried Squealer almost pleadingly, skipping from side to side and whisking his tail, 'surely there is no one among you who wants to see Jones come back?'" (Orwell 1945: 14)

flexibility, the threat of de-localization, flexible working time with tele-work, permanent staff restructuring, and changing of posts. All these elements have transformed jobs into a scarce resource. Speed becomes adaptation – "flexibility," in the new economic jargon. Changes are also occurring at a quick pace in universities.

> Universities are no longer a quiet place to teach, make academic work with a slow rhythm and contemplate the universe as it happened centuries ago. Now universities are powerful, complex, demanding and competitive business that requires continuous and large scale investments (Agencia Española de la Calidad, ANECA 2009).

Such business-oriented, speed management jargon has penetrated everywhere. This is another use of propaganda, based in globalization, in which the university is touted as a site of efficiency, adaptation, competitiveness, and so on. It is the triumph of managerialism over education. Education is one of the promising spaces to expand business opportunities, hence the rise of for-profit universities that promise fast degrees. Speed becomes an instant readaptation to a changing environment, a reality that is difficult to foresee.

Automation, automatic response, reflexes instead of reflection: This is one of the issues Ellul thought about. In a technological system, immediate response is the appropriate action. The system works as a whole and decisions are instantaneous. In fact, one of the effects of these automatizations has to do with the ability to focus on tasks, messages, and information. Cyber-intellectual labor becomes a hyper-activity dealing with information – receiving and producing, recombining and adapting, closing the circle of information. The abstraction affects labor; as Ellul would say, workers themselves vanish. This abstraction is what allows people to use machines such as computers in so many different ways. Doctors, engineers, architects, media producers, writers, and so on all base their work on the same machine, but with completely different purposes. What they have in common is to sell segments of their time. Those segments are reunited and recombined to produce something new. It is as if people, the workers themselves, are becoming redundant. What has value is the fragment of time the worker sells as the increasing necessity of nonstop connection illustrates. Mobile phones, mobile Internet connectors and other devices facilitate the need to be constantly on-line and constantly available to sell one's time. The workday begins with the first access to e-mail or mobile call. This is why the workplace, schedules, and labor time all lose their meaning.

Franco Berardi articulates a paradox: There are more and more unemployed people but personal time devoted to labor grows with each generation. According to Berardi the calculus of work hours in 1935 was 95,000 h in a lifetime. In 1972 it was 40,000; but in the 2000s we are approaching 100,000 h. Recent proposals in Europe to move the retirement age to 67 help to further illuminate this point.

Informational hyper-stimulus leads to the identification of another scarce resource: Even attention can be treated as another economic factor. Attention-deficit-disorder is a growing mental condition among young people, perhaps owing to this technological situation. Approximately 9.5 % or 5.4 million children 4–17 years of age have been diagnosed with ADHD as of 2007 in the United States. This may illustrate how:

> The colonization of time has been a fundamental issue in the modern history of capitalist development: the anthropological mutation that capitalism produced in the human mind

> and in daily life has, above all, transformed the perception of time. But we are now leaping into the unknown – digital technologies have enabled absolute acceleration, and the short-circuiting of attention time. As info-workers are exposed to a growing mass of stimuli that cannot be dealt with according to the intensive modalities of pleasure and knowledge, acceleration leads to an impoverishment of experience. More information, less meaning. More information, less pleasure (Berardi 2010).

4 Computers and Work

Thanks to automation, the nature of labor and work has fundamentally changed, as has already been suggested. But more can be said. As Italian Marxist Berardi argues, in a system for exchanging information, the format of exchange is crucial, as a selective or marginalizing element. If someone tries to transmit signals with a different format than what is used to program the network, the signals become meaningless and therefore inefficient. On one level, the Internet is a medium for erasing signs of individuality or to de-personalize – while, on another, on-line discourse affirms individuality: YouTube, Facebook, Myspace, et al. (One paramount witness to this you-philosophy was the *Time* magazine declaration of the "Person of the Year" for 2006 as "You – Yes, You!")

What is the function of this paradoxical new you-ness in cyberspace? According to Ellul in the *Technological System*:

> Man cannot live and work in a technological society unless he receives a certain number of complementary satisfactions allowing him to overcome the drawbacks. Spare-time activities, distractions, their organization, are not superfluous; they cannot be done away with for the sake of something more useful; they do not represent a true rise in the standard of living. They are thoroughly indispensable in making up for the uninteresting work, the deculturation caused by specialization, the nervous tension due to the excessive speed of all operations, the acceleration of progress requiring difficult readjustment. All these things, which are brought on by technological development, can be tolerated only if man finds a new level of compensations (Ellul 1980: 62).

Is you-ness a kind of ersatz compensation offered to balance a pervasive laboring that invades every aspect of life?

Technology in general makes work more productive or more efficient, to evoke an economic mantra. Wealth accumulates because different technologies either make the usual resources more productive or they create new avenues for the extracting of value. Behind the idea of privatization is the drive to open up new opportunities for business (Gates et al. 1995). At the same time, financial benefit is not the only motivator for work. For many engineers, programmers, and media artists the main purpose is to produce exciting new artifacts or to put imagination in motion. This motivation to do good work is associated with the beginning of the Internet boom, when the idea of being brilliant as well as making money were both entrepreneurial incentives. The "old days" were filled with amazing stories about how a bright idea could lead to success. Remember Yahoo, Netscape, and many other start-ups based at once in imagination and financial risk taking.

But these dreams ended abruptly. Interestingly enough, the idea that cognitarians should rebel against the economic system was repeated in different places. Pekka Himanen's *The Hacker Ethic and the Spirit of the Information Age* (2001), for instance, initiated movement toward a better understanding of labor, work, activism, and possibilities open not only by technologies but also through new ways to understand the politics of technology. Free software and free knowledge organizing tried to offer a counter-account of computer technology. The effort at first was not to try to find a *Temporarily Autonomous Zone*, as proposed by Hakim Bey (2003), but to use technology to isolate certain points. The effort focused on recruiting engineers, programmers, and computer scientists to a movement in favor of liberating or redirecting the aims of technological development. Hackers could form a new social class because in information technology they are those who offer tools and means to keep up with the state of affairs in which we all live. Two basic classes were distinguished: those who produced information and those who wanted to own or control it. Hackers wanted to make all information freely available to everyone so that social relations could be based on such foundations as confidence, collaboration, reputation, and a gift economy.

In 2004 a famous manifesto, echoing Marx, Wark proclaimed:

> Hackers create the possibility of new things entering the world. Not always great things, or even good things, but new things. In art, in science, in philosophy and culture, in any production of knowledge where data can be gathered, where information can be extracted from it, and where in that information new possibilities for the world produced, there are hackers hacking the new out of the old. And yet while we create these new worlds, we do not possess them. That which we create is mortgaged to others, and to the interests of others, to states and corporations who control the means for making worlds we alone discover. We do not own what we produce – it owns us (Wark 2004: 4).

But as Berardi, in contrast, pointed out,

> with the dotcom crash, cognitive labor has separated itself from capital. Digital artisans, who felt like entrepreneurs of their own labor during the 1990s, are slowly realizing that they have been deceived, expropriated, and this will create the conditions for a new consciousness of cognitive workers. The latter will realize that despite having all the productive power, they have been expropriated of its fruits by a minority of ignorant speculators who are only good at handling the legal and financial aspects of the productive process. The unproductive section of the virtual class, the lawyers and the accountants, appropriate the cognitive surplus value of physicists and engineers, of chemists, writers and media operators. But they can detach themselves from the juridical and financial castle of semio-capitalism, and build a direct relation with society, with the users: then maybe the process of the autonomous self-organization of cognitive labor will begin. This process is already under way, as the experiences of media activism and the creation of networks of solidarity from migrant labor show (Berardi 2009: 80).

The disillusion of many was presented in an obscure movie, *August* (2009), which portrayed the dot.com crash. A young computer engineer becomes a successful CEO but only for 2 years, the time it takes the Internet bubble to burst. Asked to give a talk before his colleagues, he claims that what drove computer workers to start all those little companies with venture capital funds was not just the prospect of economic success but the apparently endless possibilities that computer technology opened up. The illusion of making something real, the ability of using the imagination to invent, innovate, and create something completely new was the strongest motive for work, much more than money. But he did not realize that sooner or later there would be a takeover by the people with the money, the so-called "market."

All discourse about the liberating power of the Internet devolves into a new kind of business. Disillusion with this state of affairs – with echoes of Ellul's criticisms – can be found even among high-tech experts. Berardi with Geert Lovink, a member of the Institute of Network Culture, claim:

> Bankers and investors are not the real decision makers, they are participants in an economy of gestural confusion. The real process of predatory power has become automated. The transfer of resources and wealth from those who produce to those who do nothing except oversee the abstract patterns of financial transactions is embedded in the machine, in the software that governs the machine. Forget about governments and party politics. Those puppets who pretend to be leaders are talking nonsense. The paternalistic options they offer around 'austerity measures' underscore a rampant cynicism internal to party politics: they all know they lost the power to model finance capitalism years ago. Needless to say, the political class [is] anxious to perform the act of control and sacrifice social resources of the future in the form of budget cuts in order to 'satisfy the markets.' Stop listening to them, stop voting for them, stop hoping and cursing them. They are just pimps, and politics is dead (Berardi and Lovink 2011).

It is not difficult to appreciate the autonomy of the machine, of the technological system, stated here.

5 Conclusion: Is There Any Hope?

Issues such as global climate change, environmental pollution, financial meltdowns, and other challenges threaten a precarious level of wellbeing in post-industrial societies. Insofar as the present is dominated by a kind of techno-capitalism it is difficult to imagine alternatives. So many voices are simply calling for more of the same policies that have prevailed in public discourse for the last 20 years: more flexibility, greater liberalization, increased privatization, enhanced productivity, and so on. But surely we cannot simply accept things as they are – a situation in which Ellul provides a model for engaged, scholarly response.

Indeed, decades ago Ellul analyzed how money – the making of which is the defining goal of technocapitalism – is an abstraction that crowds out ethical reflection. The contemporary version of the work ethic (i.e., that work is virtuous insofar as it produces monetary wealth) subordinates being to having. Counter ways of life require us to de-emphasize money, give economic activity a smaller role, slow technological progress, and bring the personal and spiritual life back to the forefront of consideration. Under such changed circumstances, money would no longer cause global, collective or social problems. It would no longer be necessary to take sides on economic theory or to join a system (Ellul 1984).

Indeed, there now exist multiple voices that extend the spirit of Ellul by calling for de-growth as a realistic option to avoid disaster. The idea that growth is killing the well being of the world – not just in ecological terms but in economic ones as well – is beginning to be taken seriously. De-growth theoreticians (Latouche 2004) argue that we should reject the mantra of liberal economy that sees annual increases of in GNP as the only way to happiness and to invest instead in protecting what has become one of the most scarce resources: calm enjoyment of the ever present richness of the natural world and the practices of friendship. We could then start talking about "a-growthism," as in "a-theism." After all, rejecting the current economic orthodoxy means abandoning a faith system, a religion (Nelson 2010).

To this end we need doggedly and rigorously to deconstruct the phenomenon of development – a process initiated decades ago by Wolfgang Sachs and associates (1991) but that desperately needs continued. The term "development" has been redefined and qualified so much that it has become meaningless. Yet despite its failings, this magical word continues to command devotion across the political spectrum. Contributing to a necessary reassessment of development idolatry are contemporary hacker communities, free knowledge and peer-to-peer groups, as well as activists on the commons, all of whom promote the reusing, recycling, and saving of resources. Members of these groups not naïve; they know the temptations of technology and how difficult is to enact true freedom in cyberspace. But Ellul himself offers some hope:

> *The* [technological] *system exists in all its rigor, but it exists within the society, living in and off the society and grafted upon it. There is a duality here exactly as there is between nature and the machine. The machine works because of natural products, but it does not transform nature into a machine. Society too is a 'natural product.' At a certain level, culture and nature overlap, forming society, in a totality that becomes a nature for man. And into this complex comes a foreign body, intrusive and irreplaceable: the technological system. It does not turn society into a machine* (Ellul 1980: 18, italics in the original).

Insofar as this is true there is a possibility of contestation because society is a strange body that short circuits technology. These short circuits are what hackers and cyber-activists are looking for because it represents the possibility of freedom.

Acknowledgements This paper was possible thanks to the national project "Science, technology and society: A multi-linear analysis on communities of knowledge and action in cyberspace," and Hidden Innovation: Change of Paradigm in Innovation Studies, FFI2011-25475 sponsored by the Spanish Ministry for Science and Innovation, code FFI 2009-07709.

References

Aiestaran, Ignacio. 2010. La res digitans y el nacimiento de la wikisofía en las redes tecnológicas de la información. *Ontology Studies* 10: 257–267.

Alonso, Andoni, and Iñaki Arzoz. 2003. *La Nueva Ciudad de Dios. Un Ensayo Cibercultural sobre el Tecnohermetismo*. Madrid: Siruela.

ANECA. 2009. *El Debate sobre las Competencias*. Madrid: Ministerio de Ciencia y Educación.

Armitage, John (ed.). 2000. *From modernism to hypermodernism and beyond*. London: Sage.

Armitage, John (ed.). 2011. *Virilio now: Current perspectives in virilio studies*. Cambridge: Polity Press.

Baudrillard, Jean. 1988. *The ecstasy of communication*. New York: Semiotext(e).
Berardi, Franco. 2005. What does cognitariat mean? *Cultural Studies Review* 11(2): 57–63.
Berardi, Franco. 2009. *Precarious rhapsody. Semio-capitalism and the pathologies of the post-alpha generation*. London: Autonomedia.
Berardi, Franco. 2010. Cognitarian subjectivation E-Flux. http://www.e-flux.com/journal/cognitarian-subjectivation/. Accessed 10 Dec 2011.
Berardi, Franco, and Geert Lovink. 2011. A call to the army of love and to the army of software. http://networkcultures.org/wpmu/geert/2011/10/12/franco-berardi-geert-lovink-a-call-to-the-army-of-love-and-to-the-army-of-software/. Accessed 10 Dec 2011.
Bey, Hakim. 2003. *TAZ: The temporary autonomous zone, ontological anarchy, poetic terrorism*. Brooklyn: Autonomedia.
Bloom, Harold. 1997. *Omens of the millenium: The gnosis of angels, dreams, and resurrection*. New York: Riverhead Books.
de Teilhard Chardin, Pierre. 1959. *L'avenir de l'homme*. Paris: Éditions du Seuil.
Eco, Umberto. 1994. *Apocalypse postponed*. Bloomington: Indiana University Press.
Egan, Greg. 1994. *Permutation city*. London: Millenium Orion.
Ellul, Jacques. 1962. The technological order. *Technology and Culture* 3(4): 394–421.
Ellul, Jacques. 1973. *Propaganda: The formation of men's attitudes*. Trans. Konrad Kellen and Jean Lerner. New York: Random House.
Ellul, Jacques. 1980. *The technological system*. Trans. J. Neugroschel. New York: Continuum.
Ellul, Jacques. 1984. *Money and power*. Westmont: Inter-Varsity Press.
Florensa Giménez, Albert. 2010. *La Vida Humana en el Medi Tècnic: El Pensament de Jacques Ellul*. Barcelona: Claret.
Gates, Bill, Nathan Myhrvold, and Peter Rinearson. 1995. *The road ahead*. New York: Viking Penguin.
Gilles, Jim. 2010. Briefing: Computer traders blamed for wall street crash. *New scientist*. http://www.newscientist.com/article/dn18900-briefing-computer-traders-blamed-for-wall-street-crash.html. Accessed 10 Dec 2011.
Himanen, Pekka. 2001. *The hacker ethic and the spirit of the information age*. New York: Random.
Illich, Ivan. 1974. *Tools for conviviality*. London: Marion Boyars.
Klein, Naomi. 2008. *The shock doctrine*. New York: Metropolitan Books.
Kurzweil, Ray. 2000. *The age of spiritual machines: When computers exceed human intelligence*. New York: Penguin.
Latouche, Serge. 2004. Why less should be much more. *Monde Diplomatique*. www.mondediplo.com/degrowth/latouche. Accessed 10 Dec 2011.
Leinweber, David J. 2009. *Nerds on wall street: Math, machines and wired markets*. New Haven: John Wiley.
Lèvy, Pierre. 1994. *L'intelligence collective. Pour une anthropologie du cyberespace*. Paris: La Découverte.
Luhby, Tami. 2011. Who are the 1 percent?, *CNN Money*. http://money.cnn.com/2011/10/20/news/economy/occupy_wall_street_income/index.htm. Accessed 10 Dec 2011.
Nelson, Robert H. 2010. *The new holy wars: Economic religion versus environmental religion in contemporary America*. University Park: Pennsylvania State University Press, with the Independent Institute.
Noble, David. 1999. *The religion of technology: The divinity of man and the spirit of invention*. New York: Penguin.
Orwell, George. 1945. *Animal farm: A fairy story*. London: Secker & Warburg.
Sachs, Wolfgang (ed.). 1991. *The development dictionary: A guild to knowledge as power*. London: Zed Books.
Serra, Sanchís, and D. Arturo. 2009. *La Técnica según Jacques Ellul*. Barcelona: Liberlibro.Com.
Serra, Sanchís, and D. Arturo. 2011. *Introducción al Pensamiento Social de Jacques Ellul*. Barcelona: Aebius.
Stelarc. 2011. Projects. http://stelarc.org/?catID=20247. Accessed 10 Dec 2011.

Stephenson, Neal. 2000. *Snow crash*. New York: Spectra.
Stork, David G. 1996. *HAL's legacy: 2001's computer as dream and reality*. Cambridge, MA: MIT Press.
Stross, Charles. 2003. *Singularity sky*. London: Ace Books.
Tipler, Frank. 1997. *The physics of immortality: Modern cosmology, god and the resurrection of the dead*. New York: Anchor.
Virilio, Paul. 1977. *Speed and politics: An essay on dromology*. New York: Semiotext(e).
Wark, Mackenzie. 2004. *A hacker manifesto*. Cambridge, MA: Harvard University Press.
Wertheimer, Margareth. 1999. *The pearly gates of cyberspace*. London: Virago.

Movies

Chick, Austin, dir. 2009. *August*. 57th & Irving Productions, Original Media, Periscope Entertainment.
Ferguson, Charles, dir. 2010. *Inside job*. Sony Pictures Classics.
Niccol, Andrew, dir. 1997. G*attaca*. Columbia Pictures.

Chapter 10
Fukushima: A Tsunami of Technological Order

José Luís Garcia and Helena Mateus Jerónimo

Reflecting on the nuclear accidents at Three Mile Island and Chernobyl, in *Le Bluff technologique* (1988: 109), Jacques Ellul examines the paradox of increased unpredictability linked to technological power defined in terms of efficiency. Modern technological progress brings with it the desire to control nature and tame chance by means of calculating rationality that reduces contingencies, yet contemporary technological society has increasingly been confronted with incalculable complexities and become vulnerable to unexpected threats. Far from disappearing, as modernity claimed, unpredictability has become endemic as a result of the prodigious multiplication and power of our means of action.

Catastrophes such as that which struck Japan early 2011, as a result of the accident in the Fukushima nuclear power plant, the most serious accident ever in the history of nuclear power plants after Chernobyl, demolish claims to extremely small risk probabilities for complex systems and upend the delicate balance between costs and benefits argued by safety experts. The Fukushima disaster demonstrated once again that in contemporary societies, vulnerabilities and threats are difficult to locate or predict, being both incalculable and impossible to offset. But what the Japan disaster showed even more forcefully was that those vulnerabilities and threats derive from an infinite number of contingencies, brought about by either natural or technological events, which may interact to create potentially destructive systems, with the case of nuclear power energy production as an extreme example. Fukushima reproduces on a large scale a strong trend in disasters in the techno-dependent

J.L. Garcia (✉)
Institute of Social Sciences of the University of Lisbon (ICS-UL),
Av. Prof. Aníbal de Bettencourt, 9, 1600-189 Lisbon, Portugal
e-mail: jlgarcia@ics.ul.pt

H.M. Jerónimo
School of Economics and Management of the Technical University
of Lisbon (ISEG-UTL) & SOCIUS, Rua Miguel Lupi 20, 1249-078 Lisbon, Portugal

H.M. Jerónimo et al. (eds.), *Jacques Ellul and the Technological Society in the 21st Century*, Philosophy of Engineering and Technology 13,
DOI 10.1007/978-94-007-6658-7_10, © Springer Science+Business Media Dordrecht 2013

societies of today: natural disasters tend to have calamitous consequences with different levels of human interaction. The natural tsunami that ravaged Japan was also a technological tsunami with ecological, social, economic, and political consequences. The Fukushima disaster may be thought of not just as a disaster for Japan, but for the technological order.

The arguments of those who have argued we should be far more prudent in our approach to the technical power that humanity seeks to exercise over nature and itself – by writers who include, along with Ellul, Günther Anders, Hannah Arendt, and Hans Jonas – are valuable contributions to dealing with the difficult and dangerous scenarios of our time. Our analysis of the March 11th disaster begins in Sect. 2 with a simple narrative showing how a natural disaster in a country at the forefront of technological development precipitated an accident in a nuclear power plant that in turn led to a chain of calamities at multiple levels. The devastation caused by the tsunami was followed by terrible consequences because of a technological system that has been regarded as a major scientific and technical achievement. Fukushima demonstrates that it is not only in technologically less sophisticated or less well regulated countries that nuclear accidents become catastrophic events. Fukushima shows that an accident in a technical system on which there is significant dependence easily provokes a chain reaction in other systems supporting human life.

How did we become constructors of a world with such catastrophic potential? How is it possible to continue to insist on this form of energy? Section 3 responds by considering how the nuclear threat is played down as a result of the euphoric notion that human vulnerability can gradually be overcome by the increasing ability of science, technology and probability analysis to control and predict events. A serious nuclear accident is commonly said to be a very low-probability risk, thus affirming an "all-under-control" mentality, with the management of contingencies restricted to statistical studies, risk communications, and related technoscientific operations. This is why there are enormous resistances to admitting we may lose control of technological systems at the same time they produce new uncertainties and dangers. The real historical situation of the modern world embodies a logic that reconciles such a mentality with current forms of economic organization, based mainly on monetary profit and loss and the interests of the major market economy industries. This allows economic organizations to weaken regulatory procedures in the name of job creation or other political power benefits.[1]

The final substantive Sect. 4 retrieves Ellul's idea of "foresightedness" (*prévoyance*) as the basis of a political and social approach that can take on not only the uncertainties of the world, but also those generated by technical systems, in order to illuminate our choices and decisions. When faced with calamities and damages that

[1] Economic organizations do not so much "bypass" regulations as convince governments to weaken them in the name of interests that appeal to voters more directly than the delayed benefits that regulations realize. In the case of nuclear power, the real driver has been government prestige, energy independence of foreign oil (certainly in Japan and France), and actually pressures from some segments of the environmental movement to reduce carbon emissions.

appear to arise out of the blue but are, in the final analysis, the outcomes of our technological systems, their interactions, and our dependencies, foresightedness emerges as a response both rational and virtuous, however difficult.

1 Japan: The Vulnerability of a Country at the Forefront of Technical Progress

On March 11th, 2011, a few hours after the northeastern coast had been powerfully shaken by an earthquake measuring 9.0 on the Richter scale and then swallowed up by a tsunami that rushed violently inland, Japan sounded the nuclear alarm. The Fukushima I Daiichi power plant, one of the largest in the world, has six reactors, and had been in operation for 40 years. The power plant security systems had responded as programmed to the earthquake, automatically switching off the electricity supply; but a 14-m wave destroyed the refrigeration systems, which pump in the water required to cool the nuclear fuel rods. Fires and explosions then occurred in four of the reactors. The uncontrolled release of radiation as a result of the explosions contaminated air, water, and ground, and forced the evacuation of 80,000 people who lived within a radius of 20 km of the power plant. Over a wider area, within a radius of 20–30 km, several thousand more residents were advised to stay home, not to open doors or windows, not to switch on air conditioners, and not to consume locally produced food, in which levels of radiation were found above those deemed acceptable. As a preventive measure, the authorities distributed thousands of doses of iodine to help protect the thyroid against the effects of radiation. One month after the catastrophe and many hundreds of aftershocks, the level of the accident was raised from five to seven on the International Nuclear and Radiological Event Scale, placing it in the same category as Chernobyl. The severity of the accident reached the point of threatening the Tokyo metropolitan area, some 250 km north from the power plant and the most populous in the world, with some 37 million inhabitants.[2]

The consequences of the Fukushima disaster have continued to affect the day-to-day Japanese life. First, the earthquake and tsunami destroyed in short order all that humans had laboriously built or mastered over a period of decades, and much of the associated economic prosperity and technological achievements. In addition, many people's dreams and efforts were reduced to a sea of destruction. The provisional total loss of human life is estimated at 26,000 dead and missing. Then the chaotic situation was made worse by the nuclear threat, implying increased

[2]For a more detailed narrative, see Ribault and Ribault (2012). For a record that oscillates between personal experience and reportage, see Ferrier (2012) and Vollmann (2011). Three (unidentified) writers of the Chernobyl generation have put together articles and documents on the Fukushima disaster and published them under the name of Arkadi Filine (2012), one of the 800,000 "liquidators" of Chernobyl.

uncertainty and anxiety. Today Tomioka, Futaba, and Okuma, in the no-entry zone within a radius of 20 km from the plant, are ghost towns, their inhabitants dispossessed of their homes and belongings. At the entrance to Futaba, a sign welcoming visitors stated the following: "Nuclear Power: The Energy for a Better Future," giving an intimation of the promises of abundant, cheap, clean, and safe power.

Residents will not be able to return for decades. The accident in the power plant left an area of 2,400 km^2 in need of decontamination and 29 million cubic meters of radioactive soil, according to the Japanese government's preliminary estimates. The government also projects that the plant may remain closed and that the work of decontamination may last for at least 40 years. The list of food products in which traces of radioactivity are being found has been growing, and includes meat, rice, and powdered milk (*The Guardian*, 18.07.2011; 18.11.2011; *BBC News*, 6.12.2011). The politicians responsible for regulation of the nuclear industry and of Japan's energy colossus Tepco (*Tokyo Electric Power Company*), which operates the Fukushima power plant, failed to maintain the social trust they demanded of citizens. Even though it had been known since 2002 that Tepco had concealed a series of incidents in its nuclear power plants and falsified several reports prepared for the Nuclear Safety Agency (Lambert 2011: 6), the company played down the dangers posed by meldowns at the plant. The serious failures of regulation are a betrayal of the citizenry and have greatly increased the climate of fear.[3] Tepco did not fulfil its duties competently and responsibly, and those with political responsibility failed to monitor the situation and respond adequately.

Work has continued inside the power plant, over a year after the accident, although levels of radioactivity inside the reactors remained lethal. In November 2011 a group of 30 journalists (four from the international media) were escorted into the nuclear complex to report on activities being carried out in the damaged buildings (*The Guardian*, 12.11.2011). Until then strenuous efforts had been made to recover the cooling pumps, which in the days following the tsunami caused explosions, fires, and leaks of radioactivity. From the first moment of the disaster, teams of Tepco technicians labored in shifts trying to cool the fuel chambers. They became known as "nuclear samurais" or "the heroes of Fukushima," because they were working under extremely dangerous conditions and in circumstances that may well have irreversible consequences for their lives. The dilemma at the time was as follows: In order to reduce the temperature in the reactors and thus avoid an even greater disaster, Tepco had to spray large quantities of water on the reactors and on the pools of spent nuclear fuel, and this ended up flooding the buildings and underground galleries with radioactive water. This led to leakages of radioactive water, which could not have been avoided without reducing the volume of water being used, which would have allowed temperatures to rise and risk explosion. Consequently, Tepco had to carry out discharges of radioactive water into the Pacific Ocean in order to empty out the reservoirs, spreading contamination that harmed the fishing industry. Not until December 2011, 9 months after the accident, did the

[3] Tepco was saved from bankruptcy by the government itself, with an injection of one trillion yen ($12.5bn) (*The Guardian*, 9.05.2012).

Japanese government allege that the Fukushima power plant reactors were stable, with temperatures that would not allow nuclear reactions.[4]

At a time when memories of Chernobyl were beginning to fade, and supporters of nuclear power were gaining ground in arguing for the rebirth of atomic power as a safe form of energy suited to dealing with the problem of climate change, the March 11th disaster re-ignited a debate on the potentially malign nature of nuclear. This is leading not only to the reassessment of plans to build more reactors, but also to thoughts of dismantling existing ones. Since the accident in the Ukraine, apart from a number of less serious accidents and failures, there have been six "serious incidents" and two "accidents," based on the criteria of the International Nuclear and Radiological Event Scale.[5] The fact that a nuclear accident took place in a well-organized, technologically sophisticated country like Japan, with its carefully cultivated safety image, surprised the whole world. While retaining many of the features of its traditional society, Japan today is a country in the forefront of technological development, a leader in scientific research and the export of electronics, cars, and industrial robotics, and the third largest economy in the world.[6] It is also a country with decades of experience operating nuclear reactors. Since the oil crisis in 1973 – which reduced the economic growth trajectory Japan had followed since the Second World War, with GDP growth running at an annual rate of around 10 % – the country has committed strongly to an energy policy based on nuclear power, with a view to reducing its heavy dependence on imported oil and its vulnerability arising from a lack of natural resources.

Before Fukushima, Japan's energy plans called for the design and construction of a further 14 reactors by the year 2020, thereby increasing their share of electricity production. Until now 30 % of the country's electricity came from nuclear power, enabling it to sustain the pattern of energy consumption that seems to go hand-in-hand with chrematistic get-rich economies. A year after the Fukushima disaster, in May 2012, none of the existing 54 nuclear power plants were in operation, as they had all been gradually closed for maintenance and "stress tests." For the first time since 1970, Japan was without nuclear power and was facing the prospect of a power

[4]In technical terms, the reactors are in a state of "cold shutdown," a concept which describes intact reactors with fuel cores that are in a safe and stable condition.

[5]Since 1986 there have been six "serious incidents" (Gravelines, France, 1989; Vandellos, Spain, 1989; Tokaï-Mura, Japan, 1997; Davis-Besse, USA, 2002; Paks, Hungary, 2003; and Thorp, Sellafield site, United Kingdom, 2005) and two "accidents with local consequences" (Tomsk-7, Russia, 1993; and Tokaï-Mura, Japan, 1999). Before Chernobyl, there was one "serious accident" (Maïak, Russia, 1957), two "accidents with wider consequences" (Windscale, United Kingdom, 1957; and Three Mile Island, USA, 1979), six "accidents with local consequences" (Saint-Laurent-des-Eaux, France, 1969; Lucens, Switzerland, 1969; Windscale/Sellafield, United Kingdom, 1973; Lubmin, Germany, 1975; Bohunice, Slovakia, 1977; and Saint-Laurent-des-Eaux, France, 1980) and one "serious incident" (La Hague, France, 1981) (Valin 2011: 54).

[6]Note that Japan, however, has a poor track record of space vehicle launches. Over 50 % have failed in some way. There seems to be some cultural problem in Japan with the management of technology. India, by contrast, has an almost perfect space launch record. So does China. Note too that in the list of nuclear accidents both India and China, which have a considerable number of reactors, are conspicuous by their absence. We thank Carl Mitcham for this comment.

crunch during the long, humid summer, a time when demand for electricity is at its peak because of the intensive use of air conditioning equipment. What is remarkable is that this dramatic cutback in nuclear power (and energy consumption) has been possible. Before Fukushima, people would have said it could not be done. In this sense Fukushima has a positive side.

The complete absence of nuclear power-based electricity means that the country has to import oil and gas, which not only jeopardizes previously agreed environmental objectives, but has also given rise to the first annual trade deficit in more than three decades, and weakened the currency and therefore the balance of payments in a heavily export dependent society. These factors led to the government decision in June 2012 to restart two of the reactors at the Ohi power plant in western Japan. The announcement was made after the Prime-Minister, Yoshihiko Noda, had obtained support from the mayor of Ohi and the governor of Fukui prefecture. Anti-nuclear activists accuse the government of taking the decision too hurriedly, relegating safety concerns and residents' protests to the background, and giving in to the pressures of the nuclear industry (*The Guardian*, 16.06.2012).

Some countries with nuclear power plants, having seen once again how vulnerable the nuclear is to disasters, have decided to suspend and reassess their energy plans. The most notable case is Germany, which has a very strong anti-nuclear movement cutting right across the political spectrum. Angela Merkel, who had announced early in 2011 that she would be extending the life of nuclear power plants, did a genuine U-turn on energy policy right after the Fukushima disaster, when she decided to close 17 plants by 2022 and committed to a target of 80 % renewable electricity by 2050. In Switzerland, where 40 % of electricity comes from nuclear power, it was also decided not to extend the life of existing power plants and to cancel plans to build new ones. Belgium adopted a similar policy. In France, the country with the greatest percentage of electric power generated by nuclear, people are waiting to see whether the newly elected Hollande's promise to reduce dependence on nuclear power is going to come to fruition (*Le Point*, 11.4.2012).

The Fukushima disaster has unquestionably led to significant changes in perceptions of nuclear energy, but there are implications that go beyond simple power generation. One of the most critical aspects of progress involves enthusiastic belief that the tragic and vulnerable condition of humanity can be overcome by scientific knowledge to predict and technological power to control events. Science and technology are supposed to eliminate danger from our lives. Our inability to handle the calamities that destroy human life is an outcome of this faith.

2 A Disaster Waiting to Happen

The Japanese tragedy is like something out of "Mount Fuji in Red," one of the eight fragments in the 1990 film *Dreams* by the brilliant Japanese film-maker Akira Kurosawa. The film as a whole was a response to the reading of a piece by Fyodor Dostoyevsky wondering about the nature of dreams, which led Kurosawa to embark on a project both personal and artistic. The section at issue imagines the explosion of a nuclear power plant that initiates an eruption of Mount Fuji, one of the national symbols of Japan. It begins with images of explosions on the mountain, followed by chaos and panic and people seek to flee. The fiery, aggressive colors of the sky show how violent the explosions have been. It subsequently emerges that six reactors in a nuclear power plant have been exploding one after the other. One character, a technician from the power plant itself, explains to other survivors, a young man and a woman with two children, that the colors of the explosion represent radioactive substances: red for Plutonium-239, a very dangerous carcinogenic substance; yellow for Strontium-90, which causes leukaemia; and purple for Caesium-137, which affects reproduction, causing birth defects and deformities. "Radioactivity is invisible, and because it is dangerous they gave it colors. But that only helps you to know which type of radioactivity killed you. It is death's visiting card," he says. The mother, fearful of the threat hanging over her children, recalls that the power plant had always been described as safe, exempt from accidents and dangers. "What liars!" she exclaims. The episode ends with the young man blowing the toxic fumes away with his jacket, in a vain attempt to protect the mother and her children from the contaminated air, and the technician throwing himself into the sea, because he refuses to die a slow death.

Dreams is a mosaic of Kurosawa's reminiscences, and "Mount Fuji in Red" expresses a personal anti-nuclear position drawn from individual as well as collective Japanese cultural memories, combining the experience of earthquakes with the traumas caused by the bombardment of Tokyo and the atomic bombs of Hiroshima and Nagasaki. In "Mount Fuji in Red" there is an implicit questioning of the over-confident belief that it is possible to "domesticate" uncertainty and chance and achieve dominion over nature – a belief which serves as the foundation of attempts to establish a technological order. This fragment shows the result of a certain arrogance, which devalues uncertainty and believes that it has forever removed the risk of tragedy, but only provides a basis for their re-emergence in other horrific forms.

The denial of uncertainty and unpredictability enables the current notion of risk to prevail when weighing up the possibilities of technological accidents and disasters. The language of risk, with its scenarios of advantages and disadvantages, and measurement through probability assessments, reintroduces the notion of human vulnerability but with a suppression of uncertainty. The systematic use of statistical forecasts can be seen as a response to the modern quest for certainty, insofar as mathematics and statistics have become trademarks of true knowledge in almost every area. Forecasts always include statistical probabilities; if not, they are devalued, regarded as useless speculative exercises. But a statistical relationship is only valid in a closed system, and then only to the extent that it would be possible to ignore all factors that cannot be expressed in mathematical terms. In a world of social interaction, however, it is impossible to remove or deny unknown, random, or radically uncertain factors – that is, unpredictable events that can arise from complexity or freedom. The literature on technological accidents shows that, whether they are primarily attributable to natural causes, technological breakdowns, or human error, they always derive from multiple factors, not all of which have been precisely identified. Problems, dangers, and failures are at their most interdependent in critical situations. This does not mean underestimating the value of statistical forecasts or ignoring the experts who make them. But it is one thing to see statistical forecasts as necessary, quite another thing to see them as being practically infallible.

In statistics, many adverse possible consequences of technical systems and modern-day dangers are commonly grouped together indiscriminately as "risks," and assessment is carried out using mathematical-probabilistic techniques that convey the appearance of exactitude and of the world as more predictable, regular, and manageable than is the case. This is what enables extremely low probability estimates of a serious nuclear accident to sustain arguments for nuclear power, as Ellul (1988) properly emphasizes. Yet because they are powerful and complex technological systems, the low probabilities do not rule out the possibility of an accident, and above all the risk that such an accident may be catastrophic through the spread of radiation in the atmosphere with harmful mutational and carcinogenic effects across generations. Even if there were an extremely low probability of an accident occurring – and even more so because the design of the Fukushima power plant already incorporated automatic systems to detect and react to earthquakes, as well as defense mechanisms against tidal waves of up to 5.7 m – the effect of anything greater than these design specifications would necessarily be devastating if and when such an event did occur. Human vulnerability and the extent of the consequences also depend greatly on how the infrastructure is embedded in its context. In this particular case, the proximity to population centers, to the coast, and to farmland only accentuated the dangers. The accident thus appeared "normal" in the paradoxical acceptance of Charles Perrow, in which accidents are inherent in the infrastructure and all the ingredients for their occurrence are present. It is therefore just a matter of time (1999 [1984]: 60).

The Fukushima power plant's vulnerabilities are a good example of the situation in which, to use Ellul's language once again, "The question of possibility eclipses that of probability" (1988: 124). In a clear reference to this situation, Prime Minister Naoto Kan (who was in office at the time of the accident, but resigned in September under strong criticism for the way he had managed the crisis[7]) stated, in a conference announcing the government's commitment to renewable energy sources: "If there is a risk of accidents that could make half the land mass of our country uninhabitable, then we cannot afford to take that risk" (*The Guardian*, 8.9.2011).[8] It is true that 1 year after the accident no deaths or injuries have been directly attributed to the nuclear accident, in contrast to the many caused by the tsunami. But the experience of Chernobyl has shown that many of effects are hidden and invisible in the short term. It is not known, for example, how much radiation was released into the

[7]Yoshihiko Noda succeeded Naoto Kan. These two prime-ministers have differing attitudes toward nuclear power. While Naoto Kan became one of the strongest advocates of the abandonment of nuclear power, Yoshihiko Noda has gone ahead with a policy of reactivating reactors that have been assessed safe.

[8]One year after the accident, it was announced that the country's largest solar power station will be built in Kagoshima province, in the south-west. It will be able to supply energy to 22,000 homes.

atmosphere, particularly in the early days, because of power failures interfered with measurements; nor are the health effects of prolonged exposure to low-level radioactivity fully known. But in the words of the French Institute for Radioprotection and Nuclear Safety, the site is certainly subject to "chronic and perennial" radioactivity (*Le Monde*, 28.02.2012). Some recent studies by the Max Planck Institute for Chemistry state that the probability of contamination from severe nuclear reactor accidents is higher than expected, and Caesium-137 has extremely high dispersion rates.[9] More specifically, an article published in the U.S. *Proceedings of the National Academy of Sciences* shows that some bluefin tuna specimens caught in August 2011 on the coast of California were contaminated with radioactive substances believed to have derived from Fukushima. The levels of radiation detected are ten times higher than those found in tuna in the same area in earlier years, although still lower than those regarded as dangerous to public health. This discovery suggests that radioactivity may have been dispersed through fish (with repercussions on the food chain) more rapidly than it would have been if carried on the wind or in the water itself (*The Times*, 30.05.2012).

Despite the unknowns and the differences among experts, nuclear power plants tend to be presented as safe facilities, likely to be resilient when faced with potential problems, and with risks seen as "acceptable." This gives the impression that society is willing to accept the risks of certain activities and/or technologies in exchange for some benefits. The underlying idea is that without risks there can be no benefits, and so a reasonable "price" has to be set for this exchange. Acceptability is, however, a controversial idea, whether because most risks are imposed, and people's response is therefore one of tolerance or acquiescence rather than voluntary acceptance as such (Kasperson and Kasperson 2005 [1983]), or because it is based on the *permission* (and not *prevention*) of a certain level of risk, above which lie the supposedly really dangerous risks (Beck 1992 [1986]: 64–65; Thornton 2000), or yet again because deciding whether a probability of "one-in-a-million" is or is not safe clearly depends on certain political, social and cultural assumptions (Fischhoff et al. 1993 [1981]).

The dangers of nuclear power plants belong to the category of "unpredictable but expected effects" (*effets imprévisibles mais attendus*) and "unpredictable and unexpected effects" (*effets imprévisibles et inattendus*), to use Ellul's phrases. Prior experience, and our knowledge of nuclear power, allow us to foresee some possible consequences, even if we cannot forecast them precisely. The unpredictability lies precisely in the fact that we do not know when an accident will occur. Nuclear energy is an unknown known – we know we are running a serious danger but we unknown or deny it. "The obsession with effectiveness is so great that we take on increasingly serious risks and still hope to avoid them" (1988: 139). Perhaps this is the reason Ellul states that nuclear power "is the most striking (*frappant*) example

[9] "Probability of contamination from severe nuclear reactor accidents is higher than expected", 22 May 2012. http://www.mpic.de/Probability-of-contamination-from-severe-nuclear-reactor-accidents-is-higher-than-expected.34298.0.html?&L=2. Accessed in 25 May 2012.

of unpredictability" (1988: 175). In his view, unpredictability is one of the inbuilt features of technological progress (1988: 82). In order to understand "absolute unpredictability" – an unpredictability that cannot be remedied and is therefore a hallmark of the technical system – he makes three key points (1988: 117–120). First, we may even be able to imagine the consequences of technological developments, but not their subsequent combinations. Secondly, technical thought is incapable of envisaging technique in itself and is therefore unable to deal with its dysfunctions or adverse effects other than through the technical culture, which merely provides for the extension or improvement of that which already exists. Finally, unpredictability derives from the incommensurate growth in the quantity of risks the technical system creates. "No rational calculation can show that a timescale of five years is more or less rational than a timescale of one hundred years" (1988: 119). The technical system has become so powerful and irreversible that it has ended up losing its rationality and going beyond the realm of quantitative calculation.

Technique is also in itself ambivalent, regardless of the use to which we may put it. This is particularly true of nuclear power. Ambivalence covers a range of possibilities, from its use in medicine and energy production to war (its conversion into atomic weaponry).[10] Technique, which claims to solve so many problems, also creates others. This leads to four key propositions: all technical progress has to be paid for; it raises problems that it does not solve; it involves a large number of unpredictable effects; and the nefarious effects of technological progress cannot be divorced from the favorable. Each phase of technical progress is *destined* to solve a certain number of problems. The very movement of technique reflects convictions deeply embedded in developed countries that *everything* can be seen as a technical problem (Ellul 1988: 68). In other words, when a social, political, human, or economic problem is analyzed as a technical problem, technique becomes the perfectly suitable instrument with which to find the solution. Ellul even offers the example of the energy crisis, the "solution" to which supposedly involved building nuclear power stations (1988: 68). Ulrich Beck (2008) expressed an identical thought in an article highlighting the incoherence in wanting to solve the problems of climate change and the energy crisis through a rebirth of nuclear energy. "The incalculable dangers to which climate change is giving rise are supposed to be 'combated' with the incalculable dangers associated with nuclear power plants" (Beck 2008). In this connection, Kristin Shrader-Frechette (2011) argues along the same lines that, "there is no 'devil´s choice' between expanding nuclear fission or enduring climate change. Both are Faustian bargains. They are… a forced choice between two equally undesirable, non-exhaustive alternatives" (2011: 4).

[10]Iran's nuclear program is a case in point. Iran alleges that it is developing nuclear power for peaceful purposes, but a number of countries in the international community suspect that those purposes are of a military nature or that they could rapidly be converted to military ends.

3 Under the Gaze of Nuclear Victims

One year after the March 2011 disaster, the report released by the Nuclear Accident Independent Investigation Commission (NAIIC) seem to lend support to the idea that the prevention of catastrophe and the protection of nuclear power plants from natural dangers are calculable and are under control. Although triggered by the earthquake and tsunami, the subsequent accident at the Fukushima Plant "was a profoundly manmade disaster – that could and should have been foressen and prevented. And its effects could have been mitigated by a more effective human response", wrote Kiyoshi Kurokawa, the commission's chairman, in the report's introduction (NAIIC 2012). Once again there is a presumption that dangers are quantifiable, and a belief in scenarios which are presumed to be predictable or normal. The "stress tests" being carried out on the nuclear reactors are based on similar assumptions. The tacit consensus is that we should proceed with caution, implementing technical solutions to the problems created by technique itself, or in other words, continue along the same path. As sociologist Hermínio Martins states, this type of approach postulates that "these solutions will arrive *in time* or will emerge in a timescale which is sufficient to prevent the worst disasters" (2011: 176, italics in the original). But strictly speaking we should describe these as pseudo-solutions, or *quasi-solutions*, in the words of the American social scientist and critic Eugene Schwartz, because each one of them "engenders a residue of new technical and social problems as a result of the incompleteness deriving from the inter-relations and the limitations of closed systems, and secondary effects" (Martins 2011: 176).

Many advisers to Japan's Nuclear Safety Agency doubt the tests are reliable. They allege they fail to show the plants are safe, because the inspectors only check the design of the plants and assess their resistance to earthquakes and tsunamis of a given intensity on a random basis, neglecting the various anomalies that can occur in a disaster, including human error and failures in the equipment itself. Japan's nuclear safety chief, Haruki Madarame, who advises the government on these matters, denounced in Parliament the flawed nature of the country's regulations, the nuclear industry's excessive power, and a government more interested in promoting nuclear energy than in safeguarding the health of its citizens (*The New York Times*, 15.02.2012). He backed up his statements with various examples: officials did not give serious consideration to what would happen if electric power were lost at a nuclear station, because they believed that Japan's power grid was far more reliable than those in other countries; officials gave little attention to new studies raising the possibility of large earthquakes off the coast. Even the prime-minister, Yoshihiko Noda, who has been in office since September 2011 and wants to reactivate nuclear reactors which pass the stress tests, was forced to acknowledge that the government, the nuclear operator, and the academic world are responsible for having placed too much faith in the "myth of safety" of nuclear power. "We can no longer make the excuse that what happened was unpredictable and outside our imagination" (*The Guardian*, 3.03.2012; 9.03.2012).

Nuclear power is the highest expression of the dangers of modern technology, given that it is incalculable and impossible to offset, that responsibility for it cannot be imputed to anyone, and that its repercussions extend to generations as yet unborn. So it is difficult to understand why we continue to be governed by the misguided idea that scientific rationality, probabilistic calculation and the technological approach are sufficient for confronting its dangers, and how we avoid thinking about ecological, social and economic alternatives for energy policy. The fact that Fukushima succumbed to a 14-m tsunami, when it was prepared only for one of 6 m or less (the calculations were carried out in the 1970s), tends to be explained as a failure of risk analysis or of the company itself. In fact, in addition to the facts outlined by Haruki Madarame, we know that Tepco ignored the scenario (which it described as being "unrealistic") set out in a 2008 internal report that forecast the possibility of a 10.2-m tsunami and recommended the protecting wall to be raised.

Plans to build and extend the life of nuclear plants are strongly economical. Generally speaking these plants are located near the sea because, despite salt corrosion, seawater is a free resource for a plant that needs a great deal of water. In Japan's case, power plants are unusual in that between five and seven reactors are concentrated in the same space, whereas the country's vulnerability to earthquakes and tsunamis would counsel against this. An article in *Nature* explains that the concentration of reactors in one place means they are all exposed to the same dangers (terrorist attacks, as well as earthquakes and tsunamis) and that radiation emissions from one reactor may hinder the recovery of the others (Macilwain 2011). Japan was accused in particular of allowing a certain promiscuity of interests between Tepco and the Nuclear and Industrial Safety Agency (NISA), which in turn is an organ of the Ministry of Economics, Trade and Industry (METI). Indeed one of the measures adopted after Fukushima in order to boost residents' trust has been to place NISA under the Environmental Ministry. The politically irresponsible nature of the original situation was highlighted by Japanese philosopher Kenichi Mishima, professor at the University of Tokyo Keizai: "For me, the Fukushima disaster is the result of the failure of democratic control over industrial technology. Building reactors in an earthquake zone is a crime under the law, a form of organized terrorism against our citizens" (2011: 44).

When we are unable to predict events – as happened with the Japanese experts, who were unable to predict the violence and magnitude of the natural elements and therefore underestimated them – Ellul argues for the need to show "foresightedness," accepting that "the worst" has "always become the possible," that the worst has become "probable, not as a function of the probability calculus, which is worthless in these circumstances, but because high risks are effectively piled on top of each other" (1988: 124). Any technology has advantages and negative effects: it cannot be deconstructed so as to retain the former and avoid the latter. Ellul (1954: 92) argues, moreover, that the atomic bomb was not the outcome of the perversity of technicians, but an indispensable stage of atomic research. This trajectory of positive and negative effects is the essence of that inherent growth that is typical of technique. For Ellul, we must imagine the worst and exercise foresight.

Accordingly we should think of the uncertainties produced by complex, high-impact technological systems as producers of disasters that are certain to occur. In this context, it is misleading to use the concept of risk in the face of a superlative and uncontrollable power, which should be envisaged as embodying the certainty of future catastrophes.

Theoretical critics of technology are not the only ones who emphasize the dangers of nuclear power plants. The Japanese writer Kenzaburo Oé (2011), winner of the 1994 Nobel Prize for Literature, in an interview with *Le Monde* shortly after the Fukushima disaster, remarked: "A nuclear disaster seems a distant and unlikely possibility, but it is always with us." Oé, whose *Hiroshima Notes* (1997 [1965]) represent a comprehensive literary project to gather together the testimonies of survivors of the atomic bombings, also contended that there is a clear link between the bombing and nuclear power plants – in the careless disregard for human life – and that this is the worst betrayal of the memory of the victims of Hiroshima. In metaphorical terms, he believes we are being observed in our decisions and postures on nuclear power by those who died in the atomic bombing. "Japan has entered a new phase, but once again we are under the gaze of nuclear victims, those men and women who displayed great courage in their suffering. The lessons we may learn from the disaster will depend on the firm resolution of those who live on not to repeat the same mistakes." Those who lived through the experience of the atomic bombings cannot consider nuclear energy only in terms of industrial productivity or as a recipe for growth. The experience of nuclear disasters must be engraved in human memory, even more than natural disasters, because they are the work of humans themselves, Oé argues.

The nuclear dangers that threaten humanity highlight broader issues in the form of the constraints imposed by a vast bureaucratic machine and the enormous growth of industrial technologies. The power of this combination is greater than any ability to understand it, and leads us to act without being fully aware of what we are doing. Simple interventions in complex systems often have complex rather than simple results. Serious nuclear accidents are an eloquent example: They are events of great magnitude triggered by such simple things that it is difficult to form a mental picture of what might happen. This is the reason why it is a mistake to try to distinguish between the peaceful and bellicose uses of atomic power; all of us, without knowing it, may be implicated directly or indirectly in actions with unforeseeable effects. This in turn has profoundly altered our moral condition. This idea belongs to the technological thinker Günther Anders, who devoted much time and attention to the dangers of nuclear power. It embodies the notion that modern technology brought with it the possibility that we may be innocently guilty. Understanding the implications of technology in terms of a new moral responsibility means acknowledging that we are the precursors of a new kind of moral guilt. In Anders's words,

> after Chernobyl, inasmuch as no one can now feign ignorance, the defenders of nuclear power are consciously committing a crime. The crime carries not only the label of 'genocide' – what a use that is for the adverb 'only' – but 'globocide,' destruction of the planet. Those in favor of nuclear power, as well as those who support nuclear waste treatment and reprocessing plants, are no better than President Truman when he ordered the bombing of

> Hiroshima. In fact they are worse than he is, because people today know a lot more than the ingenuous president could have known in his day. They know what they are doing; he did not. That we mortals should perish because of a nuclear missile or on account of a supposedly peaceful power plant, amounts to exactly the same thing in the end (Anders 2006 [1981]: 317–318).

Michaël Ferrier, a professor of literature who lives in Tokyo and experienced first hand the tragedy of Japan, offers testimony that reinforces Anders's perspective. Ferrier writes that we have "legalized and normalized putting our lives in danger, we have accepted the unacceptable," "we have become used to living an amputated life… so that the nuclear machine may carry on as if nothing had happened, based on the pretext that the effects will not be visible, and will be scientifically debatable a few years hence – enough time to absorb the poison – and that the situation has all the appearance of 'normality'" (2012: 248). In his view, if Fukushima rhymes unfortunately with Hiroshima, it is not because of phonetic repetition or the two events are outcomes of atomic energy. While Hiroshima was something specific and extreme, an act of war that sought to decimate a city and subdue a people, Fukushima is the prototype of a certain technological and economic system, sold to us as being safe, and one which made an opulent lifestyle possible (2012: 255).

In this new condition, the best way of promoting awareness of the possibility of catastrophes arising from human technical power is to believe that they may indeed occur. In dealing with threats such as nuclear power the appropriate approach does not just use known factors to calculate probabilities but above all considers unknowns and how these might affect technological actions. Where the consequences of technological action are not clear, unknown aspects must be recognized and included even when they cannot be investigated and understood in great depth. This prudent way of thinking should be a systematic and integral procedure of every ethical assessment of technology, and in many instances should be imperative.

This approach echoes the heuristics of fear promoted by Hans Jonas (1984), who defends the instrumental priority of adverse forecasts over optimistic ones. The idea of threat – whether physical, existential, spiritual, or natural – should ground the assessment of the unknown risks that hover over us. As a method to avoid the dangers we face, anticipating scenarios in which human and natural life is destroyed may be crucial to preserve life. In the words of Jean-Pierre Dupuy (2008), for whom Jonas's legacy is a fundamental point of reference, we must think of nuclear accidents as embodying a dual contradiction: on the one hand, they are "instruments of fate," inasmuch as they are a necessary condition for the occurrence of a nuclear catastrophe; on the other, they are the "opposite of fate," in that accidents may not happen. Invoking the fate of a nuclear calamity as a result of a possible nuclear accident does not mean that one is fatalistic, but expresses rather an attitude of extreme prudence. This is a form of self-imposed zeal, of believing that fate is possible, of believing that the impossible is indeed possible, so that it will never occur.

4 Conclusion

The Fukushima disaster is an event that can take us further in the discussion of technical power which, as Ellul has argued, tends to be autonomous in relation to forms of democratic control, political regulation, environmental values, and the conditions of human life itself. It magnifies uncertainty and brings us face to face with a new vision of the tragic. The March 2011 disaster began with a dramatic combination of two factors: on the one hand, the frequent incidence of earthquakes in Japan and, on the other, the threat that nuclear power plants represent. While the first is of natural origin, and something that Japan has lived with throughout its history, the second is anthropogenic, the product of twentieth century science and technology. This combination had damaging, cumulative consequences, which multiplied in a chain of unfortunate events. The technical accident, which had one origin in a natural calamity, made it glaringly obvious that, in contemporary societies, where individual and social life takes place largely in the context of and in relationships of dependency on densely tangled networks of interdependent technological means, many of the vulnerabilities and threats are not easy to locate or predict. Potentially, they are everywhere, because they derive from the many contingencies inherent in systems potentially as destructive as the production of energy through nuclear power plants. Other obviously related threats can be pointed to with regard to the emergence of infectious diseases than can cause pandemics in a globally interconnected world, the fragilities of cyber-network dependencies, and carbon-based energy production leading to global climate change.

The thesis that the potential dangers of nuclear energy or any other advanced technological activity should be accepted is based on a fallacious argument. First, because the practice of and measuring risk by calculations of probability fails to cover all the unpredictabilities along with our ignorance. The risks of nuclear power plants are above all simply uncertainties. Secondly, because the gains of nuclear power and many other technological actions are realized under the burden of possible harms that are so destructive, severe, and irreversible as to be unacceptable. It is a situation in which gains in terms of production are achieved in the expensive shadow of potentially whole-scale catastrophe.

We may better understand the intellectual project of thinkers such as Ellul and others by realizing that their interest in understanding the technological context of the modern world is, above all, an inquiry into our loss of the ability to regulate the complexity of the technological system and the difficulty of recognizing our own ignorance and uncertainties. But the more problems that arise as a result of that technological system, the more we look to it as the only means of solving them. Then the occurrence of small or large-scale accidents, having more or less global, devastating, long-term and irreversible effects, counters the idea that everything is under control, based largely on a faith in technological effectiveness. Not only do we fail to eliminate uncertainty, but technological order actually brings us contingencies that resemble, or are even greater than, the old uncertainties generated by natural forces. Occupied as we are with the calculation of risk, we fail to take uncertainty as a given. Accepting uncertainty means taking seriously the principle of precaution in the regulation of technology.

References

Anders, Günther. 2006 [1981]. *La menace nucléaire: considérations radicales sur l'âge atomique*. Paris: Éditions Du Rocher/Le Serpent à Plumes.

Beck, Ulrich. 1992 [1986]. *Risk society: Towards a new modernity*. Trans. Mark Ritter. London: Sage.

Beck, Ulrich. 2008. All aboard the nuclear power superjet: Just don't ask about the landing strip. *The Guardian*, 17 July 2008. http://www.guardian.co.uk/commentisfree/2008/jul/17/nuclearpower.climatechange. Accessed 10 July 2012.

Dupuy, Jean-Pierre. 2008. Postface: de l'oeil du cyclone au point fixe endogène. In *Jean-Pierre Dupuy, Dans l'oeil du cyclone: Colloque de Cerisy*, ed. Mark Anspach, 297–316. Paris: Carnets Nord.

Ellul, Jacques. 1954. *La technique ou l'enjeu du siécle*. Paris: Armand Colin.

Ellul, Jacques. 1988. *Le bluff technologique*. Paris: Hachette.

Ferrier, Michaël. 2012. *Fukushima: récit d'un désastre*. Ménil-sur-l'Éstrée: Gallimard.

Filine, Arkadi. 2012. *Oublier Fukushima: texts et documents*. Quincy-sous-Sénart: Bout de la Ville.

Fischhoff, Baruch, Sarah Lichtenstein, Paul Slovic, Stephen L. Derby, and Ralph L. Keeney. 1993 [1981]. *Acceptable risk*. Cambridge: Cambridge University Press.

Jonas, Hans. 1984. *Imperative of responsibility: In search of an ethics for the technological age*. Chicago: University of Chicago Press.

Kasperson, Roger E., and Jeanne X. Kasperson. 2005 [1983]. Acceptability of human risk. In *The social contours of risk: Risk analysis, corporations and the globalisation of risk*, vol. 2, ed. Roger E. Kasperson and Jeanne X. Kasperson. London: Earthscan.

Lambert, Renaud. 2011. A Tepco e os seus accionistas. *Le Monde Diplomatique*, Portuguese Edition, 54: 6.

Macilwain, Colin. 2011. Concerns over nuclear energy are legitimate. *Nature* 471: 549.

Martins, Hermínio. 2011. *Experimentum Humanum: Civilização Tecnológica e Condição Humana* [Experimentum humanum: Technological civilization and human condition]. Lisbon: Relógio d'Água.

Mishima, Kenichi. 2011. Fukushima, un crime legal. *Philosophie Magazine, Dossier Nucléaire* 49, May, 44.

NAIIC. 2012. *The official report of Fukushima nuclear accident independent investigation commission*. The National Diet of Japan, July 2012. http://www.nirs.org/fukushima/naiic_report.pdf. Accessed 10 July 2012.

Oé, Kenzaburo. 1997 [1965]. *Hiroshima notes*. New York: Marion Boyars.

Oé, Kenzaburo. 2011. Nous sommes sous le regard des victimes. *Le Monde*, 17 March 2011, SPA8.

Perrow, Charles. 1999 [1984]. *Normal accidents: Living with high-risk technologies*. New York: Basic Books.

Ribault, Nadine, and Nadine Ribault. 2012. *Les santuaires de l'abîme: chronique du désastre de Fukushima*. Paris: Éditions de l'Encyclopédie des Nuisances.

Shrader-Frechette, Kristin. 2011. *What will work: Fighting climate change with renewable energy, not nuclear power*. Oxford: Oxford University Press.

Thornton, Joe. 2000. *Pandora's poison: Chlorine, health, and the new environmental strategy*. Cambridge, MA: MIT Press.

Valin, Murielle. 2011. Depuis 1957, vingt centrales ont connu des accidents de niveau 3 et plus. *Science & Vie*, Hors-Série *Le Choc Nucléaire*, April 2011, 54–55.

Vollmann, William T. 2011. *Fukushima: dans la zone interdite*. Auch: Éditions Tristram.

Movie

Kurosawa, Akira, dir. 1990. *Dreams*. Warner Bros, Pictures.

Chapter 11
From "the Contaminated Blood Affair" to the Mediator Scandal: Public Health, Political Responsibility, and Democracy

Patrick Troude-Chastenet

Technocratic skill is that of the expert whose general blindness envelops specialised lucidity.

Edgar Morin
Pour sortir du vingtième siècle (1981)

Does the philosophy of Jacques Ellul remain able to help us in some way understand the twenty-first century? From the September 11th attacks to the Fukushima disaster, through to the financial crisis of 2007–2008, we find a number of phenomena to illustrate his principal theories.

The September 11th attacks of 2001 went beyond anything conceivable. They signalled an increase in "exterior" violence and resulted in a decrease in "interior" liberties. The attacks also allowed developments in techniques, which, according to Ellul's views, purport to reduce all problems (for example, security on aircraft) to a technical dimension calling for a technical solution (for example, security scanning stations). "Technical solutions make worse the wrongs that they claim to address" (Ellul 1988: 118).

Since the Chernobyl nuclear accident, we should all know that the only predictable factor is the unpredictable. In light of the multidimensional Fukushima tragedy that has afflicted Japan since March 2011, we may recall Ellul's 1986 writings:

> We must start from the general observation that in our time, in the case of a serious accident, or a natural or artificial catastrophe provoked by technology, it is *never* possible to find an adequate response, whether from a technical or economic point of view… No one wants to accept the idea that technology has effectively placed us in the middle of hundreds of volcanoes (Ellul 1988: 123, italics in the original).

P. Troude-Chastenet (✉)
CMRP-Greccap, Université Montesquieu Bordeaux IV,
Avenue Léon Duguit, 33608 Cedex Pessac, France
e-mail: patrick.troude-chastenet@u-bordeaux4.fr

H.M. Jerónimo et al. (eds.), *Jacques Ellul and the Technological Society in the 21st Century*, Philosophy of Engineering and Technology 13,
DOI 10.1007/978-94-007-6658-7_11,

According to Ellul, when it comes to nuclear energy, we need to reverse the popular adage and consider the worst is *always* possible.

1 The Mediator Scandal

We now turn to an affair which, when viewed from outside France may seem anecdotal, even if involving loss of human life. The Mediator (benfluorex) drug scandal, which tarnished the reputation of French health authorities, only appeared in the public press toward the end of 2010. More than 900 complaints of manslaughter and unintentional injury were submitted to courts by patients and victim associations. The national health insurance fund (Caisse Nationale d'Assurance Maladie or CNAM) also lodged a complaint for fraud and serious deception. Several trials are in process and the government has created a public compensation fund for the victims. In September 2011, Jacques Servier, founder and owner of the pharmaceutical laboratory bearing his name, was charged with "obtaining unwarranted authorisation" and "deception regarding the properties of Mediator putting the life of people in danger" (*Le Monde,* 21.09.2011).

According to several independent pharmaco-epidemiological studies, the product could be responsible for 500–2,000 deaths,[1] taking into account those linked to valvulopathy,[2] and more than 3,500 hospitalizations. As the 36th best-selling medicine made in France in 2004, Mediator is reported to have brought in over one billion Euros since it was put on the market by Servier Laboratories in 1976. According to an enquiry carried out by the French health insurance fund CNAM, it has since cost the French social security system at least €1.2 billion. This would be split between €879 million in refunds and €315 million for medical supervision of Mediator treatment related complications. In sum, the costs subsequently incurred by health insurance companies need to be added. In France, 300,000 people were still receiving the medication on the occasion of its suspension in November 2009. Concerning the number of patients exposed to Mediator, the French health safety agency (Agence Française de Sécurité Sanitaire des Produits de Santé or AFSSAPS) currently called (Agence nationale de sécurité du médicament et des produits de santé or ANSM) revealed that the overall figure was around five million in an official document dated May 2011.[3]

---

[1]The epidemiologist Catherine Hill (at the Gustave-Roussy de Villejuif Institute) attributes a minimum of 500 deaths to Mediator. Cf. 'Number of deaths attributable to benfluorex' in *La Presse Medicale,* 2 April 2011. Agnès Fournier and Mahmoud Zureik from Inserm estimate that 1,320 died due to valvular insufficiency associated with Mediator during the period 1976–2009 (Fournier and Zureik 2012).

[2]A CNAM study revealed that the taking of Mediator by diabetic people multiplied their risk of valvulopathy (cardiac valve malfunctions) by three times (see Weill et al. 2010).

[3]However, C. Hill and M. Zureik had raised the figure to 7.2 million patients per year. IGAS Report, *La Documentation Française*, p. 130.

The Mediator scandal is generally approached from the angle of conflicts of interest in public life. The influence of powerful pharmaceutical lobbies on the public agencies whose task it is to evaluate their products would seem undeniable. However, the existence of financial links between those accountable to the health authorities and the Servier laboratories does not seem the only explanation for this scandal.

2 For Comparison: The Blood Contamination Scandal

A fruitful comparison can be made with the biggest political-health scandal of the 1980s and 1990s, namely 'the contaminated blood affair' (Favre 1992). Let us briefly examine it. To begin, it is important to remember that blood transfusion in France relies on the altruism and goodwill of donors. To give blood is a form of solidarity that has great ethical value. While blood transfusion centers were private organizations fulfilling a "public service mission," the umbrella organization was the national center for blood transfusion (Centre National de Transfusion Sanguine or CNTS). The organization followed a commercial logic as purveyor of blood products, and an administrative logic as a public service organization placed under the supervision of a health secretary who was in turn answerable to the ministry for social affairs and solidarity. Its top officials were eventually to serve prison sentences for fraud and failure to assist persons in danger.

While the AIDS virus had been isolated in 1983, and shown to be transmitted not only sexually but also through blood transfusion, Prime Minister Laurent Fabius and Social Affairs Minister Georgina Dufoix (July 1984–March 1986) were accused of delaying the introduction of systematic screening tests that would have prevented the accidental infection of several hundred hemophiliacs. The Prime Minister had indeed introduced delaying measures of a protectionist nature intended to favor Pasteur Institute's diagnostic test to the detriment of its American competitor Abbot. In Georgina Dufoix's case, saving on expenditure caused the delay. Worst of all, untreated blood products were sold over the counter until the stock ran out. These products were furthermore exposed to a heightened risk of contamination due to a lax attitude in donor selection.[4] According to the political scientist Jean Baudouin, during this period "the political and administrative authorities were the passive and often obliging receptacles of a muddle of contradictory rationalities" (Baudouin 1998: 319).

Georgina Dufoix's famous phrase[5] "responsible but not guilty" was mercilessly derided in its day. Instead, the meaning of her confession and the implicit questions she raised about the decision-making process should have been pondered.

[4]Twenty-five years on, the documentary *Sang contaminé, autopsie d'une affaire* (France 5, Philippe Pichon, 52 mn, 15 February 2011) gave a considerably more nuanced theory concerning political responsibility.

[5]Delivered on the TF1 television channel, on 3 November 1991, by the former minister after the charging of the former director of the CNTS and the former general director of health.

"Practically everyone, irrespective of rank, who was caught up in the system behaved in ways they shouldn't have." Dangers were ignored, precautions overlooked and, unbelievably, the matter treated as a merely economic issue. As risk philosopher Francois Ewald explained, "the system killed victims without any decision ever being taken" (Ewald quoted in Engel 1993: 12). Jurist Olivier Beaud demonstrated how the minister's dependence on her services increased with the technicality of the case. He also showed how during the contaminated blood crisis, criminal liability fully absorbed political responsibility and so much so that criminalizing ministerial responsibility allowed everyone to close their eyes to a seriously dysfunctional politico-administrative system: calling people to account avoided the trouble of calling institutions into question (Beaud 1999). By refusing to consider the notion of governmental crime and by substituting criminal responsibility for political accountability, political irresponsibility was institutionalized. Intending, it seems, to placate an irate public opinion looking for a scapegoat, President Mitterrand stepped in to speed up the appearance of his former ministers in court. However, they were not brought before the High Court of Justice or any suchlike body, but before an ad-hoc authority, the Republican Justice Tribunal, created in 1993, whose procedures derived from common law.

Now in representative democracies, ministers and elected officials are accountable for their acts to the parliament representing the sovereign people. Under these conditions, it is contradictory, to say the least, to affirm the principle of ministerial responsibility while simultaneously undermining it through the principle of criminal responsibility. Faced with grave accusations, the ministers tried to separate the individual from the collective and moral from penal responsibility. They also pleaded ignorance of the facts owing to the technical character of the case. They argued that all manner of screens, filters, and strangleholds within the central administration and ministerial departments meant the information reaching them was inadequate, hindering the making of right decisions. However, to consider this matter from Ellul's perspective – not so far removed from Max Weber's – the prestige attached to the political profession, and even more to that of statesmen, surely involves making a choice between several options and then assuming personal responsibility, for better or for worse (Weber 1982: 129, 2003: 149–150). Unlike the health secretary, Laurent Fabius and Georgina Dufoix were cleared of the charges brought against them, first of complicity in poisoning and later, in 1999, of the unintentional attack on the life and integrity of others. However, a constitutionalist will still consider that their political responsibility was fully engaged by reason of their position. Indeed, ministers must be held accountable for their errors of judgment. If they are in office, they must go before parliament; if out of office, to the country.

3 Again, the Mediator Scandal and Its Implications

Therefore, does the health crisis provoked by Mediator mean that the long and dramatic contaminated blood affair was of no educational value? Do not let us forget that ministers are responsible for refunding medicines and for authorizing their market launch (Commission d'Autorisation de Mise sur le Marché or AMM). The dismal performance of the Medicine Agency and, beyond it, that of the whole public health system, were such that we feel entitled to broaden the scope of our reflection and go back to some basic questions, which no regime claiming to be democratic should ignore. Beyond these cases concerning health, how and through whom are public policies determined? Who effectively makes decisions and who should take responsibility for them before the people, their representatives, or public opinion? Who is really at the helm of the state? As Denis Grison argued in his 2009 book, "[T]he golden rule of political action in all situations where uncertainty could weigh on the decision was to expect that science (speaking through the voice of experts) should remove this uncertainty before decisions were taken." Consequently, "the decision devolved on the experts and political officials merely followed their opinions" (Grison 2009: 182). However, for Grison, this ceased to be true with the contaminated blood affair and the emergence of the "precautionary principle."

An Ellulian approach would consider that, given the mass of increasingly technical decisions to make in a so-called "complex society," ministers have little option but to approve choices that have already been made by different categories of senior civil servants, technicians, and experts. These people, at the heart of the central administration and ministerial cabinets, take their legitimacy from their skills, not from universal suffrage, and exert power of a political nature in the name of techno-scientific necessities escaping all forms of control. This assumption, if it were confirmed, would signal that our societies are moving away from a traditionally democratic pattern, thus vindicating disenchanted visions of modern polyarchies. The tacit consent of public opinion would not change anything in this view. Indeed, while representative democracy is based on delegation – not abandonment – of power, it should not result in a shift from political responsibility to penal responsibility (Etien 2001), nor in the absence of control of the governed by those in power, nor in a deviation from the legal-rational model for the benefit of some sort of techno-structure.

The contaminated blood and Mediator affairs remind us of some remarks by Ellul – which no doubt deserve to be nuanced and placed in context. According to Ellul, authentic democracy "disappeared a long time ago" (Ellul 1977 [1965]: 320). The two cases in point offer an opportunity to test the relevance of his approach to the role of politics in technological societies. Its defining concept is that politics will henceforth be characterized by the reign of necessity and short-termism (Troude-Chastenet 2008).

What are the main features of this affair, and what can it teach us about the decision-making process – or lack of it – in a pluralist democracy? The first thing to note is that the exact nature of Mediator poses a problem and is even at the heart of the affair, as we have seen. Even today, even as the press harshly criticizes Servier Laboratories, it still presents Mediator as an anti-diabetic drug. Nevertheless, it is a mere adjuvant, without any real therapeutic efficiency and as it is derived from an amphetamine, it has the potential to be very dangerous as an appetite suppressant. Originally reserved for overweight diabetics, its appetite suppressing qualities made it one of the most prescribed medicines for patients who wanted to lose weight.

Despite its proven toxicity, "unidentified individuals" allowed this medicine to be sold for 33 years (1976–2009). Before it was banned, patient expenses were fully reimbursed by the French social security system. However, the firm that manufactured the drug knew about its secondary effects from the late 1960s. The firm also knew the dangers had been pointed out in American and French scientific journals and by several isolated, but persistent, practitioners.

For the benefit of those seeking to understand how things could have come to such a pass, a second point worth noting is the plethora of watchdogs that were supposed to stop the legal commercialization of harmful medicines. According to a report by the Inspection Générale des Affaires Sociales (General Inspectorate of Social Affairs or IGAS), no fewer than eight such organizations, on top of the health ministry, dealt with Mediator at some point. These were Direction Générale de la Santé (DGS); the aforementioned Agence Française de Sécurité Sanitaire des Produits de Santé (AFSSAPS) currently called ANSM and in particular its Direction de l'Evaluation (DEV); the marketing authorization agency (Commission d'Autorisation de Mise sur le Marché or AMM); la Commission de transparence (CDT), which since 2004 has been included in the Haute Autorité de Santé (High Authority for Health or HAS); la Commission Nationale de Pharmacovigilance (CNPV); le Comité Technique de Pharmacovigilance (CTPV); l'Agence Européenne des Médicaments; and finally the World Health Organization (WHO).

We might almost believe that this thicket of expert groups and organizations, in which everyone proffered advice and no one was responsible, was part of a deliberate strategy. Could the system have deliberately been set up to allow the dilution of responsibilities in case of a major problem? This view seems difficult to support and yet it almost seems calculated to vindicate those who support the pharmaco-industrial plot theory, or whoever derides the precautionary principle as an easy cop out. To put it simply, there were five protagonists or groups of protagonists involved in this affair. First, there was a pharmaceutical firm seeking profit; next, the experts who were or were not part of the government agencies that oversaw the manufacturing chain and clarified issues for decision makers; then, the ministers themselves, who were supposed to take charge of public policies; then the prescribing physicians, with a minority of whistleblowers among them; and, finally, the patients who were or would become victims of the drug.

Despite the presence of a so-called commission for transparency overseeing key actors, an impression of complexity and opacity emanates from this case. The story began at the end of the 1950s, when pharmaceutical firms were attempting to dissociate the hunger-inhibiting effect of amphetamines from their potentially dangerous stimulant effects. Researchers at Servier would thus correspondingly test a certain number of amphetamine by-product molecules. Among these were norfenfluramine and the resulting "fenfluramines" (commercialised in France under the brand names Ponderal and Isoméride) and benfluorex (under the name Mediator). All of Servier's marketing strategy consisted in presenting Mediator as an adjuvant for treating diabetes, which it may be, and not for what it was: a strong appetite suppressant (see Bensadon et al. 2011). However, as early as 1974, one study dedicated to norfenfluramine presented it – and the term was well chosen – "as a *mediator* of the actions of fenfluramine." Indeed, fenfluramine would be "officially" recognized as a dangerous substance in 1995 and would lose its AMM rating two years later. Thus, if fenfluramine (Ponderal), which acts through norfenfluramine, was deemed toxic, why was benfluorex (Mediator), which displayed the same connections, not suspected of being so?

Against all evidence to the contrary, the Servier group would defend the (incorrect) theory according to which: (a) Fenfluramine is part of a distinct family of amphetamines which has appetite suppressant qualities without having other side effects and (b) that benfluorex is not a molecule within the larger fenfluramines group but an original medicine, effective in the treatment of hypertriglyceridims[6] and also of what is termed "fatty" diabetes associated with being overweight.

In truth, benfluorex (Mediator) is just a precursor to the only active substance. "In other words," explained the authors of the IGAS report, "for thirty years (1976–2009), all patients treated with Mediator actually absorbed effective doses of norfenfluramine" (Bensadon et al. 2011: 47). How did this toxic medicine manage to escape expert vigilance and clear the series of hurdles that should have stopped its launch? These included the AMM commission, technical pharmacovigilance committees, the national pharmacovigilance commission, transparency commissions, not to mention opportunities to stop the reimbursement of listed products. Although fenfluramines were soon suspected of triggering pulmonary hypertension (PH) – a rare and fatal condition; although they had been banned from magisterial preparation by the Central Health Agency by 1995 (Direction Générale de la Santé, or DGS) together with all appetite suppressants, including benfluorex, yet, unbelievably, they continued to be sold over the counter as Mediator for another fourteen years. In one out of three cases, the drug was prescribed as an appetite suppressant. Sales also quickly doubled after the banning of another Servier medicine, Isomeride, from the same family of fenfluramines with appetite suppressant qualities and with serious side effects.[7]

The health authorities made mistake after mistake. They gave the go-ahead to a medicine belonging to a pharmacological class already proven dangerous; they authorized the maximum rate of social security reimbursement; and they finally granted it a monopoly in the lucrative appetite suppressant market while banning all its competitors precisely on public health grounds. All the while, the authorities never ceased expressing their doubts about the exact properties of Mediator, notably concerning its anti-diabetic properties and its usage as an appetite suppressant. Taking into account its "potential risk factor," benfluorex was the object of an inquiry from May 1995 onwards. However, despite similar information from medical agencies in both Italy and Spain, and Servier's decision to withdraw Mediator in

[6] An excess of triglycerides in the blood may make heart disease more likely.

[7] These side effects include pulmonary arterial hypertension – a rare and fatal disease. After prescribing Isomeride (Redux) and Pondéral (Pondimin) to hundreds and thousands of obese people in the USA and Canada as an appetite suppressant, the American laboratory responsible, under license from Servier, was convicted in 2001 and paid out $14 billion in compensation to victims.

these two countries, the case dragged on in France. Referring to a letter signed by three consultant doctors from France's health insurance system in 1997, IGAS inspectors wrote in a ministry of health report of January 2011 that "repeated alerts on the misuse of benfluorex" had not been taken into account and that "the withdrawal of Mediator should have been decided as early as 1999" – a full 10 years before the actual ban. While pointing out the "incoherence" and the "incomprehensible inactivity" of the pharmacovigilance system, the report's authors then pulled their punches when dealing with their fellow medical experts. All the while, every extra year of Mediator use brought a fresh contingent of victims. How many deaths and illnesses did these 10 years of "respite" for Servier Laboratories cause? Since 1992, several pulmonary hypertension (PH) cases associated with fenfluramines have been identified by the department of respiratory medicine at Antoine-Béclère hospital in Clamart, and reported to the ministry of health.[8] In addition, *La Presse Médicale* published a series of articles from 9 September 1992, explicitly linking the Mediator molecule to the family of fenfluramines.

All this should not have escaped the notice of the National Commission of Pharmacovigilance since the link between fenfluramines and PH cases had already been established in scientific journals such as *The British Medical Journal* (1981), *The Lancet* (15 February 1992), and *the New England Journal of Medicine* (1997), which also mentioned cases of valvulopathy. Moreover, in May 1997, *Prescrire*, an independent journal, came to the conclusion that "the continuance on the market and the support by health insurance" of Mediator deserved reconsidering. In fact, the transparency commission – an offshoot of the High Authority for Health, which was composed of government representatives, health insurance organizations, and "personalities chosen for their scientific skills" – made a recommendation on 19 November 1999. The High Authority for Health then brought out an unambiguous verdict, which reads as follows: "The level of medical service given is insufficient in light of other available medicines and therapies, to justify its support" (HAS 2006). The Minister of Labor and Solidarity, Martine Aubry, then announced the delisting of 286 medicines, including Mediator, before her cabinet decided to stagger the delisting process from the start of year 2000, in order to safeguard pharmaceutical industry employment. In 2011, one of Ms. Aubry's former advisors explained, "It is very clear that if there had been the slightest doubt about a product's safety, no minister would have been mad enough to retain support for it." There is no reason to call this statement into doubt. Successive ministers merely attained a lower sale price for Mediator, which always remained reimbursed at the highest rate through to its prohibition.

Hence, from November 1999, a group of experts at the French health and medicine watchdog AFSSAPS – an organization which employs 1,100 workers and

[8] Insofar as HTAP is a very rare pathology, its positive predictive value was strong. The treatise on pharmacology by Pr Giroud includes an article about pharmacovigilance, signed by Pr Bégaud, emphasising that in pharmacovigilance, "a single case can suffice to demonstrate the capacity of a medicine to produce a given effect" (in the IGAS report).

relies on the advice of some 2,000 experts – were confronted with a medicine whose toxicity was suspected, whose misuse was well-known, and whose inefficacy in the treatment of diabetes was widely acknowledged. However, they did not move to ban the product or take steps toward its delisting; instead, they merely ordered further studies and more reports. This was in 2005. Such were the compartmentalization of commissions and the complexity of their operations that in April 2006, in a statement before the Court of Auditors (Cour des Comptes), the agency's general director misguidedly claimed that Mediator had effectively been banned. Philippe Even[9] pithily described AFSSAPS as the perfect example of a Rube Goldberg machine when he said, "There are so many stages between identifying the victim of a medicine and the final decision that at least two or three years would be needed before cyanide was banned" *(Envoyé special,* France 2, 5.5.2011). In fact, it took another 5 years, from November 2005 through November 2009, before the drug was suspended, and almost another year before it was banned. All the while, "the benefit of the doubt went to the medicine, not to the sick person"; "with this affair, what was to blame was not an excess of the precautionary principle, but the lack thereof" (Bensadon et al. 2011). Such was the conclusion of IGAS inspectors, who also observed that "none of the general directors who succeeded each other at the head of the agency were properly informed" about the exact properties of Mediator "not at least until the end of 2010" (Bensadon et al. 2011). What then can be said of the ministers? It is more disturbing to realize that for 33 years the AFSSAPS experts were quite happy to accept deliberately garbled information provided by the pharmaceutical company, rather than taking into account existing scientific literature and/or carrying out their own studies. This would be like the police leaving the main suspect to speak freely while discounting victim statements as irrelevant in criminal cases. In spite of all this, should we accept the received opinion on the subject, which in large part suggests that this all boils down to the collusion of special interests and political cronyism?

As the IGAS inspectors have clearly established, the firm did indeed relentlessly interfere in the drug chain of surveillance and "anaesthetised" the principal players. Experts with close links to Servier Laboratories sat and voted on commissions investigating the suspected drug. Permanent pressure was indeed exerted to secure the AMM classification of Mediator as well as its market positioning and its reimbursement level. The agency responsible for regulating usage of the medicine did indeed share the same "référentiel" (Jobert and Muller 1987), the same values and mindset as the pharmaceutical industry. Beyond the Mediator "scandal," what is at issue is the omnipresence of the pharmaceutical industry: from the training of doctors, to the funding of theses, journals, and conferences. All in all, it is the health system as a whole that calls for debate. As I have said, the economic stake was considerable for the Servier firm, but so was it for successive governments who

[9]Professeur émérite à l'Université Paris-Descartes, president de l'Institut Necker and author of a report with Pr. Bernard Debré on overhauling the French system for controlling the effectiveness and safety of medications.

sought to fight unemployment. In the Ellulian approach, the question of employment protection would link the economic and the political variables in order to call into question the primacy of technical considerations. The Mediator affair involved a powerful boss, a French company, and a French molecule: unquestionably industrial patriotism, to no small extent, explains why the process of delisting a medicine considered ineffective but harmless was so slow. Tentative explanations note that Jacques Servier had chosen former left-wing and right-wing ministers as personal advisors, and that he infiltrated several Health minister cabinets. Finally, once Nicolas Sarkozy (one of the former lawyers at the firm) became President of France, he personally awarded Servier the Legion of Honor. Such explanations, however, do not go far enough.

In reality the legitimate reproach, which can be directed at various health ministers from 1999 onward, is having used the social security system as a means of supporting employment and not having deliberately attempted to poison fellow citizens – which would be economically counter-productive and politically suicidal for ministers, and even more so for an incumbent president. Need we evoke the fates of Laurent Fabius and Georgina Dufoix? The former saw his presidential ambitions ruined, and the latter suffered a permanent, possibly self-inflicted exile from political life. What politician would deliberately take the risk of becoming involved in a repeat of the contaminated blood affair and suffer the ignominy of Fabius' fate?

4 The Role of Experts

Let us now turn to the role of experts in this affair, both those at Servier and AFSAPPS, and more generally to the inner workings of the health system bureaucracy. Analysis bears out Ellul's view of the role of what he called "the offices" – a term he coined following his short but crucial experience as municipal advisor in Bordeaux (Ellul and Troude-Chastenet 2005). Overall, he argued that decision making has been appropriated by an aristocracy of technicians – his theory of the "dispossession of decision" (Troude-Chastenet 2005: 121–147). The politician has little choice but to implement decisions made upstream by teams of experts. Political action is thus hemmed in by technocratic constraints. For Ellul, as for Pierre Mendes-France, on the contrary, "to govern is to choose"; in other words, political decision making implies freedom of choice, which means the subordination of means to ends. For Ellul, the point is less about denying that an authentic political decision is possible and more about denouncing the burden of constraints weighing it down (Troude-Chastenet 1992: 55–61). From this point of view, the necessity to preserve industrial jobs does not undermine his central thesis, to the extent that the definition of technique as "seeking in everything the most effective method" is sufficiently comprehensive to encompass such a logic. In other words, the need for economic efficiency at all costs also falls within the technical logic.

It has since been established that during the contaminated blood affair, the general director of health passed information to the scientific advisors in the (political) secretary of state's cabinet, but never to the minister directly. During the Mediator affair, as I have already pointed out, the ministers were responsible for the policies involved in the reimbursement of medicine and the AMM rating. Thus the "expert" doctors who were responsible for pharmacovigilance did not inform the agency directors correctly. This is due to reasons already discussed: manipulation by experts at Servier, a conflict in interests, endogamy, corporatism, and bureaucratic inactivity in an organization making 80,000 decisions a year, to the detriment of some sensitive cases. Under these conditions, directors could not inform the Central Health Agency (DGS) which in turn could not warn the cabinet, which then could not alert the health minister. We could consequently ask ourselves if Ellul's reasoning, taken from his modest experience in the days following the Liberation of France, is still relevant today:

> I learned how small the politician's margin of action is, and how heavy the weight of bureaucracy... how unbelievably dependent he is on administrative services... As it was impossible for me, even with the most strenuous exertion, to devote serious attention to thirty files in the course of a day's work, I was forced to trust the heads of departments. Thus, trying to probe matters further, I realized several times that the dice were loaded. The files that came up to me were not based on serious scientific studies. Or else, I was merely asked to rubber-stamp conclusions I did not agree with. I finally said to myself that, if this was how bad it was for a municipal advisor, how much worse for a minister who receives not thirty, but three or four hundred files a day! We find ourselves completely dependent on the services (Ellul 1981: 50–51).

Even while the IGAS report does not entirely exonerate the successive health ministers, we return once again to the questions of decision-maker degrees of information and that of the autonomy of technique with regard to politics. It was put in these terms by Ellul in 1954: "The role of the politician is bound to diminish, as is already the case in the financial sector... bit by bit, he is deprived of his real power and reduced to a 'show' role" (Ellul 1981: 237) – as in the story of the emperor with no clothes. One can even feel anger or compassion, according to one's nature, as in an interview with Roselyne Bachelot, who was health minister from May 2007 to November 2010. Having recalled that doctors prescribed Mediator wrongly, she argued that new software was needed to help in making prescriptions.[10] Thus, technical solutions were offered as a remedy to human failure. Failing to take into account the supposed origin of the problem (collusion with industry, collective irresponsibility and/or bureaucratic paralysis), the response is a technical solution: The fallible human will be "corrected" by infallible software.

According to the Weberian concept, which is supposed to inspire our democratic regimes, the political leader assumes "exclusive personal responsibility while the civil servant faithfully executes his orders" (Weber 1982 [1917, 1919]: 128–129). His neutrality is the very condition for his lack of political accountability, while on the contrary, the three qualities of an authentic political figure are "passion (passionate dedication to a cause), a sense of responsibility, and the ability to glance" (Weber 1982: 162). Does the Mediator affair make irrelevant the Weberian

[10] Interview on 15.12.2010 on RTL taken from the press.

distinction between a leader who has sovereign authority over political goals and a civil servant who mobilizes technical means? First of all, ministers depend on two types of aides: scientific and political advisors, who can both be found at the heart of central administrations as well as within their cabinets. This means that in many cases, it is the technicians (senior civil servants or otherwise) who, due to technical imperatives, shape decisions of a political nature for which political leaders are made to assume responsibility. The distinction between civil servant and political leader is therefore too simplistic, since it neglects a third central figure, which may appear in many guises, whether or not a member of the administration: the expert. Of the expert Ellul wrote, in a book completed in October 1986, "he used to be the guardian of scientific objectivity, he was independent of conflicting interests. He was supposed to 'say what is right on the basis of what is true'" (Ellul 1988: 230). Now, the public has seen that the expert takes sides in power games, is attached to an organization, and has defended this organization in court (Ellul 1988: 230). Have the experts who were responsible for pharmacovigilance not proven a perfect illustration of this analysis? To this day, only the AFSSAPS's director has submitted his resignation, while none of the eight ministers potentially involved in the case were investigated. Should we be outraged in the name of a retributive conception of justice? As long as leaders remain under the influence of technocrats and blindly follow their opinions without considering their contents, we can only expect new affairs in which scapegoats serve only to mask structural problems.

Various authors have since argued that the contaminated blood affair meant the rationality of decision-making could no longer be satisfied through cost/benefit analysis, because the costs were, in general, unknown. A distinction had therefore to be maintained between the (technical) opinion of the expert and the (ethical) decision of the elected official. The latter could never be exonerated from a decision which is and must remain that of the elected official (Ewald 1997: 123). But does this not confuse what *is* with what *should be*? Of course, the minister's political responsibility, in its very nature, extends to decisions taken by others. A minister is also equally responsible either for action or inaction. Politicians should acquaint themselves with the facts and then take decisions. However, for politicians to demonstrate their understanding and exert their "glance," as Weber terms it, they should at least have something to look at. They should have reliable information on which to base their decisions. And such information, to the best of our knowledge, has not been present throughout the Mediator affair.

Rather than increasing the level of protection for patients, expert bodies and control procedures were multiplied, thus diluting responsibilities. Thus it is not by strengthening health bureaucracies that we can achieve better protection against risks. The precautionary principle, often invoked in unessential matters, has been glaring in its absence in essential ones. It has been said that the pupil should be placed at the center of the education system. The same might be said of the patient who should also be placed at the heart of the public health system; only here, words should be supplemented by actions. Patients are the primary concern and their voices must no longer be stifled; channels must be devised, in the form of associations, through which their voices can be heard, individually and collectively. The right of expression must be given to people who are currently presumed incompetent, but who at least hold competence with regard to their illnesses and who have derived their own knowledge from an irreplaceable experience which no scientific knowledge can encompass. This would be one means for the expert to recognize the competence of the ordinary person. In the Mediator case, the danger lights that started flashing as early as 1999 failed to rouse the "anaesthetized" experts from their complacency. The same danger signals would undoubtedly have alerted those who were ill and subsequently accelerated the banning of this toxic product, if they had only been involved in the evaluation procedures. Moreover, the warnings came from doctors who were motivated less by militant convictions or a sense of duty than by deep concern for those in danger. The fact that through sheer determination these "few" finally prevailed over the health bureaucracy will be seen by some as proof of democratic vitality. From this point of view, this scandal is an integral part of democracy as, naturally, it rests on imperfect beings but also notably on the liberty of expression, a free press, and respect for minority opinions.

Nevertheless, if we are to address the cause and not the symptom, we need to start by rehabilitating authentic political decisions and to take notice of the economic, bureaucratic, and technoscientific determinants that weigh them down. Should we want politics to become contingent and lasting instead of necessary and short-lived, wherever possible, we need to create centers of tension able to resist all attempts at social normalization. We need to rehabilitate the virtue of resistance in the face of Leviathan and other "Mégamachines." Outside traditional political parties, through ad-hoc associations, we also need to throw a few spanners in the modern state's bureaucratic works. Only then would we exert the right of control over everything affecting our everyday life. This would include not just health, but also environmental concerns in all their different forms, planning policy, food, transportation, life-style, and local democracy. In particular, we need to start by taking the Georges Bernanos formula into consideration: "To be ready to hope that one will not be deceived, we first need to give up hope on all that can deceive us." We need to permanently sharpen our critical sense and to keep watch. To gain or preserve liberty, we need to start by taking notice of the dangers threatening it. This is, at least, the lesson of Jacques Ellul.

References

Baudouin, Jean. 1998. *Introduction à la sociologie politique*. Paris: Seuil.

Beaud, Olivier. 1999. *Le sang contaminé. Essai critique sur la criminalisation de la responsabilité des gouvernants*. Paris: PUF.

Bensadon, Anne-Carole, Etienne Marie, and Aquilino Morelle. 2011. *Enquête sur le Mediator: Synthese*. Inspection Générale des Affaires Sociales, January 2011. http://www.igas.gouv.fr/IMG/pdf/Synthese_MEDIATOR.pdf. Accessed 23 May 2012.

Ellul, Jacques. 1977 [1965]. *L'illusion politique*. Paris: Robert Laffont.

Ellul, Jacques. 1981. *À temps et à contretemps*. Interview with Madeleine Garrigou-Lagrange. Paris: Le Centurion.

Ellul, Jacques. 1988. *Le bluff technologique*. Paris: Hachette.
Ellul, Jacques, and Patrick Troude-Chastenet. 2005. *Jacques Ellul on politics, technology and Christianity*. Eugene: Wipf and Stock Publishers.
Engel, Laurence. 1993. Vers une nouvelle approche de la responsabilité. Le droit français face à la dérive américaine. *Esprit* 192(6): 5–31.
Etien, Robert. 2001. L'encadrement constitutionnel de la responsabilité de la puissance publique. In *Actes du colloque: vers de nouvelles normes en droit de la responsabilité publique*, Sénat 11 and 12 May 2001.
Ewald, François. 1997. Le retour du malin génie. Esquisse d'une philosophie de la précaution. In *Le principe de précaution dans la conduite des affaires humaines*, ed. O. Godard, 99–127. Paris: INRA/MSH.
Favre, Pierre (ed.). 1992. *Sida et politique, les premiers affrontements*. Paris: L'Harmattan.
Fournier, Agnès, and Mahmoud Zureik. 2012. Estimate of deaths due to valvular insufficiency attributable to the use of benfluorex in France. *Pharmacoepidemiology and Drug Safety* 21(4): 343–351.
Grison, Denis. 2009. *Vers une philosophie de la précaution*. Paris: L'Harmattan.
Haute Authorité de Santé (HAS). 2006. *Avis de la Commission de la Transparence*. http://www.has-sante.fr/portail/upload/docs/application/pdf/2010-12/mediator_avis_ct_mai_2006.pdf. Accessed 23 May 2012.
Jobert, Bruno, and Pierre Muller. 1987. *L'Etat en action: politiques publiques et corporatismes*. Paris: PUF.
Troude-Chastenet, Patrick. 1992. *Lire Ellul: Introduction à l'œuvre socio-politique de Jacques Ellul*. Talence: Presses Universitaires de Bordeaux.
Troude-Chastenet, Patrick. 2005. Critique de la politique et du politique dans l'œuvre de Jacques Ellul. In *Jacques Ellul, penseur sans frontières*, ed. Patrick Troude Chastenet, 121–147. Le Bouscat: L'Esprit du Temps/PUF.
Troude-Chastenet, Patrick. 2008. Jacques Ellul et la politique. In *Patrick Troude-Chastenet*, ed. La Politique, 33–59. Le Bouscat: L'Esprit du Temps/PUF.
Weber, Max. 1982 [1917, 1919]. *Le savant et le politique*. Trans. Julien Freund. Paris: Plon.
Weber, Max. 2003. *Le Savant et le politique. Une nouvelle traduction*. Trans. Catherine Colliot-Thélène. Paris: Le Découverte.
Weill, Alain, Michel Païta, Philippe Tuppin, Jean-Paul Fagot, Anke Neumann, Dominique Simon, Philippe Ricordeau, Jean-Louis Montastruc, and Huber Allemand. 2010. Benfluorex and valvular heart disease: A cohort study of a million people with diabetes mellitus. *Pharmacoepidemiolology and Drug Safety* 19(12): 1256–1262.

Chapter 12
Homo Energeticus: Technological Rationality in the Alberta Tar Sands

Nathan Kowalsky and Randolph Haluza-DeLay

The Prime Minister of Canada has described the development of Alberta's unconventional oil resources as "an enterprise of epic proportions, akin to the building of the pyramids or China's great wall, only bigger" (*Financial Post* 2006). Its proven oil reserves of 170 billion barrels are surpassed only by Saudi Arabia and Venezuela, but both Venezuela and Alberta consist mostly of the "unconventional oil" source known as oil sands or tar sands.[1]

Tar sands are composed of bitumen, a heavy viscous form of oil attached to sand and water, from which it must be extracted using heated water. Processing the sands into crude oil takes disproportionate amounts of energy, but high oil prices make it profitable. The economic impacts are so impressive that the Prime Minister has stated that Canada is an "emerging energy superpower" (*Financial Post* 2006). As discovery of new oil deposits dwindles, we rapidly approach or have passed the peak of oil production, meaning that future production will be increasingly expensive (Bardi 2009). Thus the increasing exploitation of "unconventional" oil sources such as the tar sands, shale gas, and deep sea drilling – all of which are much more expensive than sweet light crude in terms of financial capital, energy, water, manpower, and other inputs – yields a much lower Energy Return on Investment (EROI).

[1] There is a rhetorical struggle over whether to call these bitumen deposits "tar sands" – a term used since the nineteenth century – or "oil sands" – which has more recently become the official designation used by industry, government, and the mainstream media. Because the term "tar sands" is perceived to carry a *negative* connotation (e.g., Who likes "tar"?) while "oil sands" is supposed to carry a more *positive* connotation (e.g., Everyone "needs" oil), there really is no value-neutral shorthand available (Kidner 2010). We have decided to use the term "tar sands," as our view is not fundamentally at odds with a negative assessment of the petroleum status quo in Alberta.

N. Kowalsky (✉)
St. Joseph's College, University of Alberta, Edmonton, Alberta, T6G 2J5, Canada
e-mail: nek@ualberta.ca

R. Haluza-DeLay
Faculty of Social Sciences, The King's University College, 9125 50 St NW, Edmonton, Alberta, T6B 2H3, Canada
e-mail: randy.haluza-delay@kingsu.ca

H.M. Jerónimo et al. (eds.), *Jacques Ellul and the Technological Society in the 21st Century*, Philosophy of Engineering and Technology 13,
DOI 10.1007/978-94-007-6658-7_12, © Springer Science+Business Media Dordrecht 2013

At the same time, climate change has accelerated, driven by the carbon outputs of the petro-technological society. Consequently, on the one hand, petroleum-driven lifestyles have become standard, producing a technological rationale for continued petrol extraction. On the other, pessimism can result from considering how this lifestyle is threatened by the self-generated contradictions of the petroleum-fuelled treadmill of production.

Both pessimism and technological rationality evoke Jacques Ellul's critique of contemporary Western societies. Ellul's theory of technology as a social phenomenon illuminates the totalizing rhetoric of economic necessity and inevitability used to justify the particular case of the Alberta tar sands industry. Other social values – including moral goods, social stability, religion/spirituality, and environmental sustainability – are subsumed within the efficient mastery of nature, the technical value that encompasses all others. This "techno-logic" recognizes no limit but itself and thus assumes for itself a sacred character. Applying Ellul's hermeneutic to the Alberta tar sands both counters recent criticisms of technological determinism and provides a basis for the examination of tensions surrounding environmental and social concerns in Alberta.

1 Ellul's Critique of the Technological Phenomenon

For Ellul, technology (or *technique* – for ease of exposition, we shall not trouble ourselves with Ellul's fine distinctions between the two) is a societal logic, "a *Weltanschauung* that is potentially an insidious and pernicious determinant of social action" (Graham 1983: 218). This societal techno-logic can be variously described as a mindset, worldview, culture, social process, "consciousness" (Lovekin in Garrison 2010: 197), rationality, value-system, "ideology" (Garrison 2010: 197), paradigm or interpretive schema, and it produces all manner of advanced gadgets which we would not otherwise have. More importantly, however, it characterizes our so-called civilization. Technological rationality and its associated institutionalization become the total – or totalizing – milieu in which we operate (Ellul 1971: 12).

Ellul's socio-historical analysis suggests that the contemporary era comprises a profoundly new human condition (Goddard 2002). What he calls the "technical operation" is simply the fact that human beings always have had to use tools and techniques to navigate their environment (Ellul 1964: 19–20). The "technical phenomenon," by contrast, refers to the momentous unification of the theoretical sciences (*logos*) and the practical arts (*technē*) starting in the sixteenth century (Grant 1986). It is only in the context of the technical phenomenon that Francis Bacon could say "Human knowledge and human power meet in one" (*Novum Organon* 1.3). Thus, for example, when Ellul spoke to the prospect of nuclear energy, he distinguished between scientific knowledge for its own sake, and the use of that knowledge to literally gain power:

> If it were simply a matter of knowing the constitution of matter, I don't think there would be any problem… Unfortunately, in atomic research we are not dealing with knowledge, so much as manipulation, transformation and disintegration (Ellul 1982: 15–16).

Such intricate understanding of the hidden workings of the physical universe have exponentially increased our ability to redirect it toward ends of our own making. Technical logic insists on *doing* something with knowledge; "unused" or "uselessness" are synonyms for negative value in our culture, as when uncut forest is considered "lost" timber value (cf. Ellul 1982: 115–116). The technical phenomenon thus embodies the will to power, which Nietzsche saw in the history of the West: "this attitude of power and control… is the driving spirit of the technical system" (Punzo 1997: 29). *Our claim is that the Alberta tar sands instantiate this technical will to power.*

Ellul's point about technical power is that it is not evaluated according to any standards other than the limits it sets for itself. The "technical *millieu*," he says, "is formed by an accumulation of means which have established primacy over ends" (Ellul 1971: 12). Technical means become evaluated simply from the perspective of technical means – and the *only* value that guides technical operation in our society is *efficiency*. To submit technology to any other value would be to make the technology less efficient, and "everybody knows" that inefficiency is undesirable and bad.[2]

Of course, all of a society's causal operations are directed to one end or another; we do not possess petroleum fuel for its own sake, after all, but for its use – although such use is increasingly experienced as a sense of entitlement (Huber 2009). Rather, Ellul's point about techno-logic is that we do not (and do not feel the need to) submit these ends to scrutiny; what we use energy *for* is outside the realm of technoscientific and thus authoritative discourse. All ends other than technique itself are considered merely subjective or completely relative. Efficiency, by contrast, is not taken to be a subjective value; it is supposed to function the same way that "objective fact" supposedly does. Scientific rationality examines the facts and, when they have been adequately determined, deliberation ceases. The purposes of human or social activity are passed over in uncritical silence; the only thing that we can rationally deliberate about is the most efficient mode of technical operation. Thus the pursuit of power – represented by the headlong pursuit of new extraction projects in Alberta's unconventional oil sector – is a self-reinforcing cycle. By mystifying any other discourse, the appeal to "factual" efficiency for its own sake obfuscates technological discourse as rational and without legitimate alternative: it alone is objective, disinterested and value-neutral. Overall, Ellul says that:

> The term *technique*, as I use it, does not mean machines, technology, or this or that procedure for attaining an end. In our technological society, *technique* is the *totality of methods rationally arrived at and having absolute efficiency* (for a given stage of development) in *every field of human activity* (Ellul 1964: xxv).

The result is that technique comprises the limit of what is pragmatic and possible. Echoes of this logic can be heard in the rationale given for Canada's December 2011 withdrawal from the Kyoto protocol, where the Federal Environment Minister said "[withdrawing] allows us to continue to create jobs and growth in Canada."

[2]"[T]echnology does not endure any moral judgement. The technician does not tolerate any insertion of morality into his work. His work has to be free" (Ellul 1980: 145).

The minister's reference to jobs and growth only bolsters the implication that it would be inefficient for the country to depart from the petro-economic status quo.

But citizens are not supposed to be mindful of such technological rationality. For Ellul, the rejection of any values other than the increase of efficient causal power leads to the *naturalization* of the technical phenomenon. When technology functions as our environment, we take it wholly for granted – as if its existence *as is* were necessary or "second nature" (Ellul in Punzo 1997). Any uncertainty about technological hegemony is simply an inability to "face the facts." *Our contention is that the Alberta tar sands are presented as necessary, socially naturalized and thus immune to any criticism other than technological fine-tuning.*

The second effect of technique's rejection of any values other than the increase of efficient causal power is that human individuals and societies must adapt themselves to the technological system, not the other way around (Ellul 1980: 244). The requirements of efficiency call "for a social order perfectly malleable to the demands of technique, requiring that political, economic, and educational structures be constantly open to meet these demands" (Punzo 1997: 23). A technological society cannot be subject to other, heteronomous values, such as those dictated by "tradition" or non-technical "morality." The technical phenomenon *absorbs* and *integrates* such nontechnological responses, insofar as they can be fitted to the logic of the machine. According to George Graham, the "force of [Ellul's] sociology is to show that technology permeates all human activities and prepares us through education to feel at home, so to speak, in the technological system" (1983: 226). But because technique is only applicable to physical material, human beings will ultimately be treated as *nothing but physical material* by technique when they are adapted to the technical system. This is why Ellul says that technical rationality will be applied to every field of human activity. *Our claim is that the Alberta tar sands are developed such that Albertan society has to conform to the demands put on it by mining operations, and that human beings are fitted to the demands of the tar sands rather than the inverse.*

Kevin Garrison summarizes thus: Modern techno-logic is "a continual move toward rationalizing all aspects of human life, placing those aspects within a technical sphere, and destroying all possibilities for thinking or acting outside that sphere" (Garrison 2010: 197). Thus, in Pierre Bourdieu's terms, technique is what the entire social system takes for granted as the conditions of practical action ("*doxa*"), the universe of the undiscussed and undisputed (Bourdieu 1977: 168). Whenever the technical system seems to be flawed or disagreeable in some manner, the only conceivable option is to view these problems as technical problems that technical progress will eventually solve.[3] *Faith* in future technology as a salvific force reinforces the prior conviction that technology is *sacrosanct* and cannot be in any way dispensed with. *Our contention is that the Alberta tar sands are presented as*

[3]The very term "side-effect" attests to the assumption that no problem with technology is essential to technology itself.

untouchable and therefore may only be corrected by further technical remediation. It is in this way that Ellul's theory of technique amounts to the (in)famous claim that technology is *autonomous*. Ellul's sociological analyses exhibit a tension between the way the society actually is, the way it wants to be, and the way it will be if current trajectories are maintained. His point is that techno-logic is an idea of our own that nevertheless controls us. The Alberta tar sands impose demands on society that are accepted by both Albertan voters and the politicians that reiterate those demands, and thus their development is carried out as if we had no choice in the matter. *Our claim is that the Alberta tar sands are sold to Albertan society in terms of inevitability and irresistibility, when this is, in fact, not true.*

In the end, however, Ellul is not the pessimist he is made out to be by critics. In spite of places where he decries a technological tyranny, Ellul's theological writings develop the hope that he has in the face of his sociology of technology. Put simply, he declares "we must destroy the deified religious character of technique" (Ellul 2004: 89). Borrowing from his Christian background, he advocates the iconoclastic desacralization of the falsely sacrosanct technical phenomenon. But such "[i]conoclasm is possible only to the extent that one is able to give up the religious assurances of one's culture" (Ellul in Vanderburg 2004: 129), and only if those religious assurances are not already *built* into the architecture and geography of our social environment (see Huber 2009: 474). *Our contention is that the Alberta tar sands should be exposed as false gods, and only after such desacralization can truly reasonable debate take place as to how their development might proceed.*

2 Overview of Alberta's Athabasca Tar Sands

Ellul's theory of technique illuminates the particular socio-political landscape of Alberta's Athabasca tar sands, which in turn provides something of an empirical confirmation of his theory against claims that his work speaks only about technology in the abstract and thus falsely (as claimed in, e.g., Pitt 2009). As noted above, the Alberta tar sands are the world's third largest reserve, and often presented as crucial to North American continental energy security. They are considered large and secure but "dirty," with different social actors emphasizing different characteristics. As conventional oil reserves decline, it has become profitable to exploit alternative sources such as the bitumen-soaked sand and shale that make up the Alberta tar sands. Developing these reserves requires considerably more energy and water and produces more carbon emissions than developing conventional oil.[4] Tar sands development is contested, with many representations of its ecological destructiveness and social

[4]Nikiforuk (2008) provides the most accessible account of the oil sands history, production processes, relation to provincial political and economic systems, and environmental and social impacts.

consequences. In fact, it is these competing "facts" that begin to illuminate the technical rationality that Ellul highlights as central to the totalizing technological milieu.

One set of "facts" about the "oil sands," current as of early 2011:

- The estimated 170 billion barrels of recoverable bitumen "represents a massive amount of future economic activity on top of current production" (Gibbins 2010: 14).[5]
- The tar sands provide 15 % of United States crude oil imports, or 7 % of that nation's oil demand (Government of Alberta 2011).
- Production is currently approximately 1.5 billion barrels per day, estimated to be near five billion by 2020.
- The tar sands provide "huge" economic benefit according to industry, industry-oriented thinktanks, and both federal and provincial governments. These benefits include revenues, jobs, low taxes, and no sales tax. One estimate is $218 billion of capital *investment* over the next 25 years (Government of Alberta 2011).
- Aboriginal Canadians receive considerable benefit as many are employed by, or Aboriginal-managed businesses are actively engaged in, tar sands services.
- The tar sands are worthy of awe, with countless instances of innovation and technology improvements associated with this monumental task.

Finally, and most importantly, government and industry justification is "The world needs more energy and lots of it!" and "We all NEED energy" (e.g., www.albertaisenergy.ca).

Another set of "facts" are that the "tar sands":

- Result in environmental damage to watersheds, fish, and habitat for threatened species like woodland caribou.
- Carbon Emissions: "The West accounts for 53.8 % of [Canada's] GHG emissions compared to 30.4 % of Canada's population in 2007" (Gibbins 2010: 6).
- Threaten health downstream, in such communities as Fort Chipewyan.
- Impinge on Aboriginal treaty rights, and that Aboriginal participation is not as good as touted by tar sands supporters.
- Have reduced social sustainability and increased social problems in the resource-based community of Fort McMurray.
- Are not even an efficient fuel. Calculations vary but the EROI of the Alberta tar sands is approximately 3:1 (Andreoli 2010). By comparison, conventional oil in 1930 was approximately 100:1 while the estimate for conventional oil a decade ago was 20:1. At the same time, ethanol (a biofuel) is approximately 1.2:1.
- Cheat Albertans by not being governed by an effective royalty regime (Boychuk 2010).

[5]One of a myriad of such reports, *Look Before You Leap* is produced by a think-tank and tries to demonstrate the significance of the Western Canadian economy (heavily reliant on energy extraction) to the national economy, as part of an argument that the federal government should NOT regulate tar sands further, nor implement carbon emission caps. Watts (2005) asserts that think-tanks are a crucial component of the oil complex.

Industry and both the provincial and federal governments have aggressively defended the tar sands. The economic benefits of tar sands are expressed as essential to public well-being, and that "we all need energy" is presented as a *fait accompli*. Public relations campaigns, including a high profile provincial government one with a budget of $25 million, explicitly framed Albertan identity in the context of energy-production: "Energy is what makes us Albertans" and "Alberta is Energy" (Haluza-DeLay 2011). While the rapid expansion of tar sands development is contested by Aboriginal, environmental and other organizations, for the most part, they have not engaged in this discursive narrative about identity. Instead, environmental organizations have offered green economy counterproposals, emotional appeals (pictures of oil-soaked ducks, Aboriginal peoples as victims of environmental injustice), and international mobilizing to delegitimize the tar sands (Haluza-DeLay and Carter forthcoming). In other words, they have engaged in a cycle of technique that we will more fully explicate later. Meanwhile, a request for a moratorium on new tar sands approvals in 2007 signed by a number of civil society organizations and a 2009 Pastoral Letter by the Roman Catholic bishop of the region made it clear that even mainstream institutions were beginning to question the tar sands development regime.

3 Techno-Management in Action

A brief sketch of three examples should demonstrate the logic of technique in action. In each, the question arises about how our society deals with profound disagreement. While the purported answer is "science," as we can see this is complicated by deeper issues, namely, the rationality of technique, which brooks no disagreement from even positivistic science. "Proof" – that is, scientific certainty rather than "levels of confidence" – is asserted to be the measure necessary before taking action. Two of the following examples – whether tar sand development is affecting downstream human health, and whether there are significant ecological damages occurring from tar sands extraction – show the scientization of the issue, that is, they illustrate the techno-logic at work. The third example – the response to Roman Catholic Bishop Luc Bouchard's Pastoral Letter – may most clearly illustrate the imposition of technique because it illustrates the response to a social actor coming at the tar sands from a completely different angle than the dominant techno-managerial orientation.

3.1 Aboriginal Health

3.1.1 The Question: Are There Health Impacts?

The primarily Aboriginal communities downstream of the Athabasca tar sands deposits and extraction have come to believe that the extraction is causing health problems among their residents. Most prominent has been Fort Chipewyan. This community has requested health studies since the late 1990s, early in the acceleration of tar sands development. No baseline data had been collected. The concern became widespread when in 2003 physician Dr. John O'Connor publically described what he thought was an unusually high rate of rare cancers in Fort Chipewyan (Loyie 2009). O'Connor was later disciplined by Alberta College of Physicians and Surgeons for several counts of poor practice and raising "undue alarm" – all charges of which he was later cleared. Eventually, the regional health authority commissioned an independent study which, when released in November 2007, showed high levels of toxic and carcinogenic substances (such as mercury and arsenic) in fish and soil downstream of tar sands development (Timoney 2007). The provincial government countered with a health review raising counter-results. But this study raised questions about methodology and reliability and increased distrust within the communities (Brooymans 2009). A new health study in 2009 showed a statistically higher incidence of rare cancers, but the small population made conclusions contentious. Eventually, a Royal Society of Canada (2010) report asserted that there is "no feasible mechanism" for tar sands pollution to cause downstream health concerns.

With this back and forth, the public and professionals are left uncertain about what is the truth of health effects. Science is claimed as "the way" but is clearly politicized. From this respect, one might conclude that this undermines the Elullian perspective of technique as the determining orientation. But this example illustrates the scientization of politics, as much as the politicization of science. The Ft. Chipewyan community is playing the same game as the government, that is, "facts" as determined by technical experts will be the deciding factor in the moral debate. In many ways, it actually undermines their claims that colonialism, and failures of participatory process are fundamental aspects of what they consider an environmental injustice. Furthermore, the question has become "Is the health risk worth taking?" (Tenenbaum 2009), which illustrates efficiency rather than safety, well-being, or other values as the central principle.

3.2 Environmental Science

3.2.1 The Question: Are There Environmental Consequences?

A similar process has been occurring over the question of the ecological impacts of tar sands development. The context of this question is an industry-funded monitoring program called the Regional Aquatic Monitoring Program (RAMP), which is the only ongoing measure of environmental impacts. RAMP repeatedly reports no consequential environmental impacts. However, the Ayles review in 2003 determined that RAMP is scientifically inadequate. Among the problems: RAMP does not share data, presents only conclusions, has inadequate methodologies, and is run by people with inadequate credentials. In recent years, independent research has been conducted. Biostatistician Kevin Timony concluded that there was indeed ecological damage and published the findings in peer-reviewed journals (Timoney and Lee 2009, 2011). Internationally renowned University of Alberta water scientist Dr. David Schindler also concluded there were ecological consequences of tar sands developments in the first of two papers published in the prestigious journal *Proceedings of the National Academy of Sciences* (Kelly et al. 2009).

Provincial politicians dismissed such studies. Alberta ministry of environment chief scientist Dr. Preston McEachren accused Dr. Timony of "lying." McEachren's speech was reported on industry websites and widely circulated by the industry association Canadian Association of Petroleum Producers. Timony eventually won a court case by showing that he had not ignored data that McEachren had accused him of doing and McEachren – defended by taxpayer money through Alberta Environment – was ordered to offer an apology and get the accusation removed from the web. Subsequent research by Schindler (Kelly et al. 2010) received considerable media and public attention, when it also concluded that environmental damage was occurring. The response of the provincial government (through then Minister of Environment Rob Renner) was that Schindler was in the pay of environmentalists, and that government data disputes Schindler's findings. Furthering the back and forth, Schindler presented deformed fish in public while Alberta Environment scientists offered alternative explanations of the purported deformities.

At this point in Fall 2010, both Alberta Premier Ed Stelmach and the Federal Environment Minister Jim Prentice stepped in to announce "blue ribbon panels" to review environmental management of the tar sands. The Royal Society of Canada (2010) produced a report in December and the federal Oil Sands Advisory Panel (2010) released its review one week later. Both reports concluded that there is no clear scientific evidence about the degree of environmental damage, particularly since environmental monitoring has been extraordinarily poor. Industry and government touted these as proof that efforts to protect health and environment were working. The Royal Society report also noted that there is far, far less research being done (and most of the research was proprietary) on this issue than on comparable environmental controversies in the past (like acid rain deposition). The Alberta "blue ribbon panel" took months longer than anticipated. While it was intended to be a "scientific" assessment, at least one key science expert quit because the composition of the panel included political insiders and industry representatives but no environmental or Aboriginal representatives. This panel's conclusions also asserted the need for better monitoring, and once again took a scientifically cautious approach in that only effects that were demonstrated conclusively in existing technical science were validated.

As with the first example, science is claimed as "the way." We are supposed to believe as citizens that science is not political and that it simply gives us the "facts" that resolve what would otherwise be (supposedly intractable) value-laden disputes. The perception has been manufactured that *technoscience* is deciding the dispute. But by the measures of science, we can say that the science *is* being heeded. Since the science is equivocal, management operates on a cost-benefit heuristic – development proceeds and mediation of the effects *we know about* occurs in a manner that does not unnecessarily impinge on development. This is efficiency in operation. Effects *not yet proven* are not part of the technological calculus at work. Ellul's point is that efficiency is a value masquerading as a non-value.

3.3 Bishop Bouchard's Letter

3.3.1 The Question: Are There Moral Concerns?

A final example will more fully demonstrate this techno-logic, because it arrives at the tar sands issue from a very different starting point, and utilizes science but is not dominated by it. A high-profile and contentious intervention into the tar sands issue occurred early in 2009. Roman Catholic Bishop Luc Bouchard, whose Diocese of St. Paul includes the Ft. McMurray area, released an extensively researched Pastoral Letter titled *The Integrity of Creation and the Athabasca Tar sands* (Bouchard 2009; Cryderman and Loyie 2009; Abercrombie 2010). The letter consisted of four parts. The first section presented scriptural and theological reasons why safeguarding the natural environment is a religious obligation. The second summarized the environmental effects of the tar sands. The letter closed by drawing religious and moral conclusions from the analysis and then recommending individual and political actions that must be considered if the integrity of the environment is to be respected. As Bishop Bouchard concluded,

> Any one of the above destructive effects provokes moral concern, but it is when the damaging effects are all added together that the moral legitimacy of tar sands production is challenged. An even more alarming level of concern is reached when the scale of proposed future expansions… is taken into account. It is then that the full environmental threat of the tar sands and the resulting gravity of the moral issue involved is most deeply felt (Bouchard 2009: 8–9).

In an address a year later to over one hundred people at an ecumenical social justice conference in Edmonton, Bouchard said he was amazed at the quantity and tone of the responses to his letter (Warnica 2010). The most common sorts of negative responses were ad hominem references to church scandals, or statements such as "Churches should stick to morality, and what they 'know about'" and "I'll stay out of your God business if you stay out of my tar sands business." In other words, morality had no place in discussions about the tar sands. This in spite of the fact that religious faith always has to ask questions of "What is the good to be done?" and "Is what we are doing what we *should* do?" (although these are, of course, not merely questions to be asked by religiously oriented citizens).

Industry and government representatives responded to his letter, Bouchard said, "Like they were reading from the same page." They ignored the moral question as if it was already settled that tar sands development should proceed in the manner that it has. They repeatedly described their technical efforts to reduce tar sands impact. Bouchard reported that the only engagement by government or industry in the moral questions he sought to raise was by a single industry executive who pointed out in a letter to the *Calgary Herald* newspaper the social good of jobs that are created by the tar sands extraction. Bouchard pointed out that these responses indicate a prevalent societal belief that issues of economics are above moral comment and that the tar sands should only be dealt with by technical experts.

4 The Tar Sands as Technological Phenomenon

The Bouchard affair is the clearest example of the technological phenomenon at work in the tar sands. First, responses to the pastoral letter generally segregated (so-called) facts from (so-called) values, lumping religion and ethics with the latter at the end of a descending hierarchy of authority. From this perspective, the "facts" of economics are so necessary to the status quo that they cannot be subjected to any values other than their own. Therefore, the only responses to Bouchard's moral criticisms were claims about technical remediation and job creation. The technocratic managerial approach to the tar sands excludes other approaches to understanding them or taking account of their benefits and costs, thus becoming a "closed loop" unable to conceive of alternatives to the system-as-is.

Second, this approach reflects Ellul's claims about modern technology's embodiment of the will-to-power. Alberta embodies some of the democratic characteristics associated with what is variously called the "resource curse" or the "petro-state" (Adkin et al. forthcoming). Watts (2005) describes the "national organization of the oil-producing state" as being characterized by excessive elite control and a tight association with the major petrol companies (Watts 2005: 377). Despite the social and environmental costs, the Alberta government repeatedly asserted there would be "no slowdown" in approvals or production because the tar sands are considered to be the economic engine of the province.

Clearly, then, the current development of the tar sands is aimed toward economic power. What Ellul has said about nuclear power can be applied to unconventional oil:

> All atomic research is research for power. It is no longer simply 'nuclear energy.' Some, as in France, must compensate for oil deficiencies; other nations must guarantee continuous growth in energy consumption; still others must escalate nuclear weapons in order to guarantee national security. In all cases power is indeed at stake. We have the fixed idea that matter contains an unlimited amount of power and is completely at our disposal (Ellul 1982: 116).

It goes without saying that unconventional sources of oil compensate for declining conventional oil availability, and that increasing development of such sources are demanded by the ongoing growth in worldwide energy consumption. Moreover, because the Athabasca tar sands are not located in a politically volatile part of the world, they are touted as an "ethical" source of American energy security (Levant 2010; but see Hiemstra 2011). If the connection to power was not obvious enough, recall the Government of Canada's reference to becoming an energy *superpower*. Thence the relevance of Ellul's claim that "the technology of energy… is closely linked to the spirit of domination, conquest and human lust" (Ellul 1982: 116).

Our culture has trouble seeing this will-to-power as a problem. Oil (especially gasoline) gives us "command over space" through automobility (Huber 2009: 477). The EROI of petroleum has fuelled Western economic and military imperialism, sweeping away other competitor systems, such that both western culture and western economics have become the over-powerful center of a core-periphery system (Roberts and Parks 2007). Our attitude toward power is that it is a good thing (at least when used for good, and responsibly or "sustainably" developed). Progress in the efficient mastery of nature is viewed not only as proof that we are smarter and morally superior to those cultures that preceded ours (both temporally and technologically), but as actually making us smarter, morally superior, and happier. Garrison writes that "technological advance always 'bluffs' people into thinking that the world is better off than it was, primarily because… the 'great innovation' of the technical system is that the collective unconscious of technological societies are driven by an 'absolute belief in unlimited progress'" (Ellul in Garrison 2010: 201). Thus, although petro extraction is often claimed to be driven by demand, it organizes a social system that drives that demand forward, and the entire system is undergirded by the will-to-power that Ellul articulated.

In the face of this institutional momentum, there is no other option but to "face the facts." What Ellul says about nuclear power again finds relevance in this context:

> We are no longer capable of saying at any given moment, 'Enough! We're stopping!' At any given moment, we have neither the criterion nor the motivation not to pursue to the nth degree everything that can satisfy our spirit of power. [Our spirit of power] can do nothing but will the means to its own satisfaction, no matter the cost (Ellul 1982: 116).

Albertans are told that it would be economic suicide to slow down (let alone stop) tar sands development. Taxes would inevitably rise and jobs would be lost. Petrol prices would rise even higher than they are now, which means more expensive goods, less mobility, and economic stagnation. On top of this there are also home heating costs, electricity generation, and the wonderful products made from plastic to consider. In short, the purported facts are that *we cannot live without oil*: "hegemonic discourses naturalize gas consumption in practice only as an isolated consumer 'choice'; thereby reproducing the entrenched power structures embedded in the overall totality of petro-capitalism" (Huber 2009: 477).

The system of technique is thus manifested as the petrol-fuelled society. While we have made reference above to petrostates and petrocapitalism, state socialism exhibited the same expansionary "treadmill of production" (Ellul 1980: 133–145). This indicates that it is not capitalism but something deeper than either of the economic systems that dominate our epoch. We are stuck in a meta-economic system and, for the most part, do not even question it at the fundamental levels that Ellul reserved for the insidious tyranny of technique. The statement "We all NEED energy" – the position promulgated by the Canadian Association of Petroleum Producers – is a faith statement, presented as orthodoxy with criticisms or alternatives cast as heterodox. In this system, the tar sands appear as a necessary evil, and, in the Albertan consciousness, are turned into a social good.

Andrew Zimmerman describes Ellul as advancing the thesis that "large-scale technics embody some of the most insidious sources of tyranny in the modern world" (1995: 89). While this may be said to refer to the large scale of the petrostate, more importantly it refers to the technological society, its hypnotic effect on citizens, and in particular the petrol-fuelled version that is contemporary Canadian society. Zimmerman (1995) and Andrew Feenberg (2002) refer to democratization as overcoming technocratic rationality, but our analysis does not show this happening in Alberta: "It is doubtful that democratic action can be useful when the more important goals of thoughtful and reflective actions are required" (Garrison 2010: 199). Ellul would agree. Alberta appears as "a hypnotized province" (Hiemstra 2009), where even in their criticisms, civil society is still as caught by the *Weltanschauung* of technique, particularly the inability to think outside the energy status quo. There is simply no other fuel with the EROI of oil, and yet no collective discussion of reduced-energy lifestyles in an oil-constrained future. Technique thus imposes an all-encompassing structure on human action.

5 Whither a Hope?

Many have made note of a perceived pessimism in Ellul's analysis. For Winner, Ellul's analysis is "an elaborate hall of mirrors, deliberately designed to leave no passage out" (in Garrison 2010). Such observations miss the point that for Ellul, the way out is Jesus. Theological anthropology is crucial for understanding and resisting the technological society. Christianity – or more properly in Ellul's terminology, the Revelation – "prevents society from locking itself into a finished system" (Ellul in Graham 1983: 223), providing an "out" from the enclosure of technique. We understand this as both a general revelation – such as the possibility of innovative problem-solving derived from unexpected and not necessarily religious quarters – although it appears clear that Ellul thinks the special revelation of Christianity provides a more accurate standpoint from which to assess the system.

Most critics of the tar sands, even Christian ones, say a great deal about oil (accomplishing the commodity fetishism that Huber 2009 articulated). But as an iconoclastic community leader in Alberta observed, they say precious little about *love* (Wiebo Ludwig, January 2011, personal communication).[6] In Ellul's words,

> We must be quite clear that what we believe is that God's promise, received in faith, borne by us, truly changes the conditions in which we live and act. In other words, the presence of faith in Jesus Christ alters reality. We also believe that hope is in no way an escape into the

[6]Ludwig was a conservative religious leader in northern Alberta who led a self-sufficient religious commune. Dangerous sour gas wells drilled near the communal property and homes led to altercations with the rural Alberta populace and the oil and gas industry, including wellsite bombings and a gunshot death. Ludwig was eventually convicted for involvement and spent nearly 2 years in jail, becoming publically known as Alberta's first convicted eco-saboteur (Nikiforuk 2002). Ludwig died during the final editing of the present chapter, on 9 April 2012.

> future, but that it is an active force, now, and that love leads us to a deeper understanding of reality. Love is probably the most realistic possible understanding of our existence. It is not an illusion. On the contrary, it is reality itself (Ellul 2004: 87–88).

Love clearly does not make sense from the orientation of technical management, and yet it stands at the root of much religiously based analysis and Bishop Bouchard's letter in particular: "The earth, therefore, is to be treasured, loved and safeguarded" (Bouchard 2009: 4).

For Ellul, the only hope-full dialectical partner to technique was the Christian revelation (not the sociological formation of the church) (Goddard 2002) and this is illustrated in Bouchard's insertion into the debate. First, Bouchard's analysis of the tar sands did not start with the inevitability and efficiency of petrol extraction. Second, he at least implicitly desacralized technique by submitting it to a moral standard other than itself. Finally, he positioned his anthropology on other bases than humans as products of a sociotechnical regime. In this perspective, human identity/being is not first and foremost what the technological system (in our case, oil) interpellates us to be; we are to be and act out of a different identity, and this is one of the reasons the religious insertions into the debate have been met by a wild mix of responses by the public. But because the "care for creation" analysis does not fit the technocratic paradigm of resource management (either epistemically or practically), government and industry talk past it instead of to it.

Ellul's theological works are a central part of his analysis. He cannot be understood without Christian hope; and he will primarily be seen as a pessimist without this half of his analysis. This framing, and the resistance and hope borne of it, will look odd to non-Christians and nominal Christians, let alone committed Christians who do not see their faith as counter-cultural (Haluza-DeLay et al. 2013). It is clearly not consistent with the political frames of Alberta or even of the ENGOs, which merely advocate for the right sort of management for the regulators of the tar sands. The rank and file of the Alberta population – the "Martha's and Henry's," as a previous Premier used to refer to them – may find it difficult to see that even though we justify our society by appealing to technical rationality, we are where we are not because of inexorable progress but rather because of ideological and economic contingencies that we have (falsely) naturalized and (sadly) reified.

6 Conclusion

In conclusion, we have argued that the Alberta tar sands reflect pervasive technical rationality, suggesting that Ellul was right to say that technologic is the context in which modern culture exists. His analysis of technique as a system of social organization and a knowledge orientation provide traction on the intractable problems of accelerating ecological destruction. "Technique" takes the central sociological place that "capital" held in the nineteenth and first half of the twentieth century, such that technique is even more salient in the twenty-first century than capital. All the capital of the oil corporations would be insufficient to cause such social and ecological and spiritual desolation without the tools, know-how, infrastructure and manufactured social acquiescence to exhume the bituminous sand. Technique shapes the deployment of capital, to great and deleterious global effect. As David Orr has said, environmental destruction "is not the work of ignorant people. Rather it is largely the results of work by people with BAs, BSs, LLBs, MBAs, and PhDs" (Orr 1994: 7).

On our analysis, the tar sands ground Ellul's theory in the particularities of circumstances he could not have foreseen, circumventing criticisms arising from the "empirical turn" in the philosophy of technology (Achterhuis 2001). The tar sands also provide a connection between Ellul and current theorizing on petrocapitalism and automobility. Both are examples of how technique was manifested in the second half of the twentieth century and because of the decline of production after "peak oil" will change dramatically sometime in the twenty-first, providing a distinct rupture with the previous time period (Poland et al. 2011). The unfinished contestation of oil suggests a politics over petrocapitalism that is not about "oil as power objectified" but a manifestation of the colonizing of lifeworlds ever more thoroughly by technique. Ellul's "question of the century," according to Garrison, is whether or not humans will "wake up from their somnambulistic love affair with new iterations of technology and place technological growth in a dialectic tension of freedom versus one-dimensionality thinking" (Garrison 2010: 202). Given the decline of oil production and the discovery of the increasing climate costs of our carbon economy alongside the lack of government initiative and public will to move toward new energy economies, this question is truly the question of the twenty-first century.

References

Abercrombie, Sharon. 2010. Abuse of creation: Bishop joins voices challenging expanding oil sands extraction in Canada. *National Catholic Reporter*, May 14, 2010.

Achterhuis, Hans J. (ed.). 2001. *American philosophy of technology: The empirical turn.* Bloomington: Indiana University Press.

Adkin, Laurie, Byron Miller, and Naomi Krogman, eds. Forthcoming. *Political ecology and governance in Alberta.* Toronto: University of Toronto Press.

Andreoli, Derek. 2010. *Energy and unconventional oil.* Paper presented at the American Geography Association, Washington DC, April 2010.

Bardi, Ugo. 2009. Peak oil: The four stages of a new idea. *Energy* 34(3): 323–326.

Bouchard, Bishop Luc. 2009. *A pastoral letter on the integrity of creation and the Athabasca oil sands.* http://www.dioceseofstpaul.ca/index.php?option=com_docman&task=doc_download&gid=14&Itemid=5&lang=en. Accessed 8 July 2011.

Bourdieu, Pierre. 1977. *Outline of a theory of practice.* Cambridge, UK: Cambridge University Press.

Boychuk, Regan. 2010. *Misplaced generosity.* Edmonton: The Parkland Institute.

Brooymans, Hanneke. 2009. Cancer downstream from oilsands merits further study: Alta, Gov't. *Canwest News Service*, February 26, 2009.

Cryderman, Kelly, and Florence Loyie. 2009. Bishop spurns oilsands development. *Edmonton Journal* 29 January 2009: A3.

Ellul, Jacques. 1964. *The technological society.* Trans. John Wilkinson. New York: Alfred A. Knopf.

Ellul, Jacques. 1971. Ideas of technology. In *1984 and all of that*, ed. Fred H. Knelman, 11–20. Belmont: Wadsworth.

Ellul, Jacques. 1980. *The technological system*. Trans. Joachim Neugroschel. New York: Continuum.

Ellul, Jacques. 1982. A theological reflection on nuclear developments: The limits of science, technology and power. In *Waging peace: A handbook for the struggle to abolish nuclear weapons*, ed. Jim Wallis, 114–121. New York: Harper & Row.

Ellul, Jacques. 2004. *Perspectives on our age: Jacques Ellul speaks*, ed. William H. Vanderburg. 2nd ed. Trans. Joachim Neugroschel. Toronto: Anansi.

Feenberg, Andrew. 2002. *Transforming technology: A critical theory revised*. Oxford: Oxford University Press.

Financial Post. 2006. Bigger than the pyramids! *The National Post*, July 29, 2006.

Garrison, Kevin. 2010. Perpetuating the technological ideology: An Ellulian critique of Feenberg's democratized rationalization. *Bulletin of Science Technology Society* 30(3): 195–204.

Gibbins, Roger. 2010. *Look before you leap: Oil and gas, the Western Canadian economy and national prosperity*. Calgary: Canada West Foundation.

Goddard, Andrew. 2002. *Living the word, resisting the world: The life and thought of Jacques Ellul*. London: Paternoster Press.

Government of Alberta. 2011. *Facts and statistics*. Alberta Energy. http://www.energy.alberta.ca/OilSands/791.asp. Accessed 16 Dec 2011.

Graham, George. 1983. Jacques Ellul: Prophetic or apocalyptic theologian of technology? *The Political Science Reviewer* 13(1): 214–239.

Grant, George. 1986. *Technology and justice*. Toronto: Anansi.

Haluza-DeLay, Randolph B. 2011. *Energy is what makes us Albertan: Territoriality, identity and petrocapitalism in an energy-producing province*. Workshop on "Energy & Territoriality in International Perspective." Canadian Centre for German and European Studies, York University, Toronto, January 29, 2011.

Haluza-DeLay, Randolph B., and Angela Carter. Forthcoming. Social movements scaling up: Strategies and opportunities in opposing the oilsands *status quo*. In *Political ecology and governance in Alberta*: *Volume I*, ed. L. Adkin, B. Miller, and N. Krogman. Toronto: University of Toronto Press.

Haluza-DeLay, Randolph B., Michael P. Ferber, and Tim Wiebe-Neufeld. 2013. Watching avatar from 'AvaTar Sands' land. In *Avatar and nature spirituality*, ed. Bron Taylor. Waterloo: Wilfred Laurier University Press.

Hiemstra, John L. 2009. Canada's oil sands developments as icon of globalization. In *The gospel and globalization: Exploring the roots of a globalized world*, ed. M.W. Goheen and E. Glanville, 179–198. Vancouver: Regent Press.

Hiemstra, John. 2011, December 7. The masterful illusion of "Ethical Oil." *Comment Magazine*. http://www.cardus.ca/comment/article/3012/. Accessed 16 Dec 2011.

Huber, Matthew. 2009. The use of gasoline: Value, oil, and the "American way of life". *Antipode* 41(3): 465–486.

Kelly, Erin N., Jeffrey W. Short, David W. Schindler, Peter V. Hodson, Mingsheng Ma, Alvin K. Kwan, and Barbara L. Fortin. 2009. Oil sands development contributes polycyclic aromatic compounds to the Athabasca River and its tributaries. *Proceedings of the National Academy of Sciences* 106(52): 22346–22351.

Kelly, Erin N., David W. Schindler, Peter V. Hodson, Jeffrey W. Short, Roseanna Radmanovich, and Charlene C. Nielsen. 2010. Oil sands development contributes elements toxic at low concentrations to the Athabasca River and its tributaries. *Proceedings of the National Academy of Sciences* 107(37): 16178–16183.

Kidner, Keely. 2010. *Oil sands or tar sands? What these words really mean in Alberta*. Paper presented at under Western Skies: Environment, community, and culture in North America, Calgary, Canada, October 14, 2010.

Levant, Ezra. 2010. *Ethical oil: The case for Canada's oil sands*. Toronto: McClelland & Stewart.

Loyie, Florence. 2009. Doctor cleared over suggested link between cancer & oilsands. *Edmonton Journal*. http://www2.canada.com/edmontonjournal/news/cityplus/story.html?id=6951e2e4-76fc-4bd1-b32e-8a6e045be0c1. Accessed 5 Apr 2013.

Nikiforuk, Andrew. 2002. *Saboteurs: Wiebo Ludwig's war against big oil*. Toronto: Macfarlane, Walter & Ross.

Nikiforuk, Andrew. 2008. *Tar sands: Dirty oil and the future of a continent*. Vancouver: Greystone Books.

Oil Sands Advisory Panel. 2010. *A foundation for the future: Building an environmental monitoring system for the oil sands*. Ottawa: Environment Canada.

Orr, David. 1994. *Earth in mind: On education, environment, and the human prospect*. Washington, DC: Island Press.

Pitt, Joseph. 2009. The autonomy of technology. In *Technology and values: Essential readings*, ed. C. Hanks, 87–96. London: Wiley-Blackwell.

Poland, Blake, Mark Dooris, and Randolph Haluza-DeLay. 2011. Securing "supportive environments" for health in the face of ecosystem collapse: Meeting the triple threat with a sociology of creative transformation. *Health Promotion International* 26(Suppl 2): ii202–ii215.

Punzo, Vincent. 1997. Christian hope in a technological age: Jacques Ellul on the technical system and the challenge of Christian hope. *Proceedings of the American Catholic Philosophical Association* 70: 17–31.

Roberts, J.Timmons, and Bradley C. Parks. 2007. *A climate of injustice: Global inequality, North–South politics, and climate policy*. Cambridge, MA: The MIT Press.

Royal Society of Canada. 2010. *Environmental and health impacts of Canada's oil sands industry*. Edmonton: Royal Society of Canada.

Tenenbaum, David J. 2009. Oil sands development: A health risk worth taking? *Environmental Health Perspectives* 117(4): A150–A156.

Timoney, Kevin. 2007. *A study of water and sediment quality as related to public health issues*. Fort Chipewyan: Nunee Health Board Society.

Timoney, Kevin, and Peter Lee. 2009. Does the Alberta tar sands industry pollute? The scientific evidence. *The Open Conservation Biology Journal* 3: 65–81.

Timoney, Kevin P., and Peter Lee. 2011. Polycyclic aromatic hydrocarbons increase in Athabasca River Delta sediment: Temporal trends and environmental correlates. *Environmental Science and Technology* 45(10): 4278–84.

Warnica, Richard. 2010. Bishop assesses fallout from letter. *Edmonton Journal*, February 28, A5.

Watts, Michael J. 2005. Righteous oil? Human rights, the oil complex and corporate social responsibility. *Annual Review of Environment and Resources* 30(1): 373–407.

Zimmerman, Andrew D. 1995. Toward a more democratic ethic of technological governance. *Science Technology and Human Values* 20(1): 86–107.

Part III
Reason and Revelation

Chapter 13
The Reception of Jacques Ellul's Thought in French Protestantism

Frédéric Rognon

It is commonplace to observe that, during his lifetime, Jacques Ellul was better known in the United States than in France. This would seem to confirm the biblical adage according to which "a prophet is not without honor except in his own country."[1] Nevertheless, it should be pointed out that Ellul was a member of the National Synod (from 1947 to 1970) and of the National Council (from 1956 to 1968) of the *Église Réformée de France* (Reformed Church of France), respectively the legislative and executive bodies of the largest French Protestant denomination. Describing his situation in the Reformed Church, Ellul famously said that he was a part of a "minority within the Protestant minority." (Protestants represent only 2 % of the French population.) Ellul was not kind to his own community; he considered self-criticism a precondition for social criticism. At the end of his last term in office in 1971, he was awarded responsibility for drafting a report on the state of theological education to be presented at the Nancy-Pont-à-Mousson Synod of 1972 (Ellul 1972). Despite this official recognition, many of Ellul's opinions (concerning issues as diverse as the war in Algeria, Israel, Islam, apartheid in South Africa and Aides) received cold receptions, as attested by the avalanche of letters to the editor that his articles in the French Protestant weekly *Réforme* inevitably provoked. His criticism of church institutions also gave rise to many misunderstandings. Such was the case when, in *Hope in Time of Abandonment*, he proclaimed that the Holy Spirit had left the Reformed Church overly preoccupied with financial and real estate issues, as if the Holy Spirit was still present, "we would know about it" (Ellul 2004a). Or alternatively again when, on the eve of the regional Synod in autumn 1983, he predicted,

[1] Mt 13:57.

F. Rognon (✉)
Faculté de Théologie Protestante, University of Strasbourg
Palais Universitaire, Bureau 49, Strasbourg, France
e-mail: rognonfr@yahoo.fr

H.M. Jerónimo et al. (eds.), *Jacques Ellul and the Technological Society in the 21st Century*, Philosophy of Engineering and Technology 13,
DOI 10.1007/978-94-007-6658-7_13,

in line with the Congregationalist ecclesiology he embraced at the end of this life, that the Reformed Church in France would disappear in less than 10 years were the Synods and all regional and national bodies not abolished with their authority returned to the local communities (Ellul 1983).

Jacques Ellul made no qualms about voicing his deep disappointment over Christian conventionality, beginning with Protestants who he maintained behaved as if desperately seeking to strip themselves of their distinctiveness. His frustration was all the greater as he had placed great hopes in the Reformed Church, which he felt could become an authentic "revolutionary movement," by which he meant a vector of Christian presence in the modern world capable of questioning the mythologies and idolatries of our time and of promoting a way of life radically different from conventional norms (Ellul 2007). In other words, it was a force that could bring about the only "necessary revolution" (Ellul 1969) capable of challenging devotion to the state and subservience to the gods of technological society. This hope inspired his failed attempt to convene a general assembly of French Protestantism (Ellul 1985a, b, c, 1986a). The reality of life within the Reformed Church in France fell so far short of Ellul's expectations that he subsequently turned his interest to small local communities, a process thus mirroring his short-lived political career. Ellul's involvement in the French Reformed Church began at the national level in 1947, following a brief experience in municipal politics in his hometown of Bordeaux in 1944 and, although it was long in coming, his disillusionment with the institutional church echoed his discovery 30 years earlier of the political illusion (Ellul 2004b).

Seventeen years after his death, we can now begin to perceive a substantial interest in Ellul's sociology and his criticism of the technological society beyond the French Protestant microcosm, amongst neo-communalists, ecologists, and anti-globalization activists. But what about his theological, biblical, and ethical writings? How are these aspects of his work, which he always considered inseparable from his social criticism, seen today? To answer this question, we first describe Ellul's position within the theological and ecclesiastical landscape of contemporary French Protestantism. Second, we describe some of the current French Ellulian theologians.

1 Jacques Ellul Within the Theological and Ecclesiastical Landscape of Contemporary French Protestantism

The French theological landscape has changed significantly in the last 20 years. Not only does the theology of Karl Barth (Ellul 1979, 1986a, b; Chastenet 1994) that inspired Ellul now seem obsolete, but the same is also true of the Third World and political theologies which he fought for (Ellul 1963, 2006). Theological liberalism and social Christianity are staging a comeback and some of their spokespeople take paradoxically similar positions to those held by Ellul. Although they consider his relationship to Scripture to be too orthodox or too confessional, the liberals admire the centrality of freedom (Ellul 1973, 1984) in Ellulian ethics (Guivarch 2008). As for social Christians, although they object to the disjunction between history and the Kingdom that is at the heart of Ellul's theology, they applaud his engagement in social issues, such as his work to establish anti-delinquency centers (Ellul and Charrier 1971). Evangelicals appreciate Ellul's bible-centrism and quote him in support of the conservative positions they take on bioethics and sexual ethics, including homosexuality. They also approve of his Judeo-Christian exclusivism and his distrust of interfaith dialogue. They do not however accept Ellul's soteriology and balk at his criticism of technological compliance within the church, positions they judge too eccentric. It should however be noted that evangelicalism is an extremely heterogeneous movement and the boundaries between its factions are more porous and fluctuating than ever.

Turning to the field of theological journals, we see that *Foi et Vie*, of which Ellul was editor from 1969 to 1986, and then editor emeritus until his death in 1994, is still being published, although it no longer takes a narrowly Barthian stance nor does it seek to promote the early Barth, which Ellul used to counter leftist Barthians and disciples of the later Barth (Ellul 1986b). Although the journal *Autres Temps* went out of circulation a few years ago, social Christianity has not totally disappeared from the scene; indeed, interest in it revives with each successive economic or social crisis (Lavignotte 2010). Theological liberalism is represented by two journals, *Théolib* and *Évangile et Liberté*, which no longer indulge in the dogmatic infighting that characterized them in the past. The *Revue d'Histoire et de Philosophie Religieuses* edited by the Lutheran school of theology in Strasbourg and the *Études Théologiques et Religieuses* published by the Reformed schools of theology in Paris and Montpellier favor articles written in the liberal tradition, but are not sectarian in their approach. The *Revue réformée* of Calvinist persuasion and the evangelical journal *Hokhma* continue to give voice to their respective orthodoxies, but no longer trade anathemas as they did during Ellul's lifetime. It seems as if the relentless theological battles that were once the national pastime of French Protestants, Ellul foremost among them, have lost their appeal. Times have changed.

Today, French Protestantism more closely resembles a kaleidoscope of sensibilities, perpetually reconfiguring itself in response to institutional change. The Lutheran and Reformed churches of Alsace and Moselle have united to form the Union of Protestant Churches in Alsace and Lorraine (UEPAL in French) without actually merging. Beyond Alsace-Lorraine, the Reformed Church of France and the Evangelical Lutheran Church of France are preparing a merger which will result in the founding in 2013, of the United Protestant Church of France (EPUdF in French). In recent years, the Protestant Federation of France (in French, FPF) has increasingly welcomed evangelicals and is currently presided over by the Free Church leader Claude Baty. Other evangelicals who judge the venerable Federation too liberal have countered by creating their own alliance, the National Council of Evangelicals in France (CNEF in French). While some evangelical churches have chosen sides, others have opted for dual affiliation in an effort to accommodate both parties. In this largely pacified theological and ecclesiastical landscape, the sharply defined well-nigh peremptory positions that Ellul advocated might seem out of place but they continue to appeal to those who deplore the absence of real debate between proponents of conflicting or irreconcilable positions.

Not only theologians but ministers and laypeople offer a sympathetic ear to the words of Ellul whose books are now back in print after a 10-year lapse following his death. The sociological essays were the first to be reedited, creating a new interest in his theological works (Ellul 2006, 2007). French Protestants are now rediscovering Ellul's biblical commentaries and his writings on the ethics of freedom. This new-found interest can be attributed, on the one hand, to the importance of the topic of Creation in church debate and, on the other, to the struggle against idolatry, which is a leitmotif in the French Calvinist tradition. It is worth noting that even today Ellul is omnipresent in *Réforme*, the French Protestant weekly to which he contributed a total of 218 articles. It is not unusual for both authors of the feature "Disputatio," in which two thinkers confront their respective opinions on social, political or theological questions, to quote Ellul in support of their opposing positions! (Lavignotte and Vaquin 2010; Rognon 2011; Moreau 2011) Since *Réforme* so closely reflects the state of opinion within French Protestantism, the significance of these frequent references to Ellul cannot be underestimated.

Moreover, all the major Protestant media and schools of theology have at least one person on their staff or faculty who has been influenced by Ellul's thought. For the schools of theology: Oliver Abel[2] in Paris, Jean-François Zorn[3] in Montpellier, Yannick Imbert[4] in Aix-en-Provence, Neal Blough[5] and Jean-Claude Girondin[6] Vaux-sur-Seine and yours truly[7] in Strasbourg, not to mention Bernard Rordorf[8] in Geneva. Media figures include Jean-Sébastien Ingrand, director of the Protestant Multimedia Library in Strasbourg, Antoine Nouis (Nouis 2002, 2003, 2006, 2007, 2010, 2011), bestselling author published in Geneva by *Labor et Fides* and in Lyon by *Olivétan*. In addition, many Ellulians write for *Réforme*: its director Antoine Nouis has a weekly column,[9] while Olivier Abel, Stéphane Lavignotte, Jean-François Hérouard and yours truly are regular contributors. Finally, signs of a new generation of dynamic Ellulians can be seen in the "Bible and Creation" network of Protestant ecologists founded by Stéphane Lavignotte.[10]

To summarize, many distinguished theologians and ministers of different generations refer explicitly to Ellul, among them: Jacques Maury (born in 1920), Gabriel Vahanian (1927), Michel Leplay (1927), Bernard Rordorf (1940), Jean Baubérot (1941), Jean-François Hérouard (1943), Jean-François Zorn (1946),

[2] Professor of Philosophical Ethics.

[3] Professor of Church History.

[4] Professor of Apologetics.

[5] Professor of History.

[6] Associate Professor of Sociology.

[7] Professor of Philosophy.

[8] Emeritus Professor of Systematic Theology.

[9] Since January 2012, Antoine Nouis has been the editorial director of *Réforme*.

[10] See http://blog.bibleetcreation.com/

Michel Rodes (1947), Olivier Abel (1953), Sylvain Dujancourt (1954), Antoine Nouis (1955), Jean-Sebastian Ingrand (1969), Stéphane Lavignotte (1970). Analysis of the existential journeys and intellectual itineraries of each of these thinkers would prove useful for evaluating and understanding the impact of the thought of Ellul on the French Protestant landscape today, and perhaps tomorrow (Rognon 2012).

2 Portraits of Some Current French Ellulian Theologians

Jacques Maury was parish minister, president of the *Église Réformée de France* (Reformed Church of France) from 1968 to 1977, and president of the *Fédération Protestante de France* (Protestant Federation of France, which federates the most important protestant Churches) from 1977 to 1987. He is a precious witness to relations between Ellul and his Church: relations of fervent engagement and strong inner criticism. He remembers how Jacques Ellul showed total solidarity with his Church, when after 1968 people expressed new aspirations, and when many ministers resigned. Jacques Maury disagreed with Jacques Ellul about Israel and South Africa, but he was strongly influenced by him on the revolutionary mission of the Church in the society: Jacques Ellul taught him not to adhere to superficial revolutionary slogans, but to see and show the real and profound problems. Jacques Maury sees in Jacques Ellul an actual revolutionary man and much more so than many so-called revolutionaries, and took him as the example of how to be revolutionary inside of Church.

Gabriel Vahanian first taught in the United States for a number of years, at Syracuse University, and then in Strasbourg at the *Faculté de Théologie Protestante* (Faculty of Protestant Theology), where he retired in 1995. His theology is deeply indebted to the thought of Ellul, with whom he engaged in a forthright dialogue over such fundamental questions as technology, utopia, eschatology and hermeneutics (Vahanian 1961, 1976, 1983, 1992, 2008). Both men attended the annual seminars organized by Enrico Castelli in Rome. Still active and productive, Gabriel Vahanian is highly regarded in both the French-speaking and English-speaking worlds. Although not an unconditional Ellulian, he frequently refers to Ellul in his writings, acknowledging a debt or setting out a difference. An architect of the critical reception of Ellul, he considers criticism to be a form of homage to the intellectual heritage bequeathed by the author of *The Technological Society*.

Michel Leplay has held several ministries: minister in parish, president of the *Commission des Ministères* (Commission of Ministries, which recruits new ministers), director of the protestant weekly *Réforme*, member of the *Groupe des Dombes* (crew of ecumenical dialogue), vice-president of *Amitié Judéo-Chrétienne de France* (AJCF, Jews-Christians Friendship of France). Michel Leplay is a man of dialogue, especially with the Catholic Church and with Judaism (Leplay 2002, 2006). It is within this last context that he met Ellul. They collaborated on several occasions, when the professor of Bordeaux wrote in *Réforme*, and above all in the activities of the AJCF. Friendship with Jewish people is very important for Michel

Leplay, and the fight against Anti-Semitism was a common engagement for both men. However, the cause of dialogue with everybody did prevent Michel Leplay from following Ellul in his vigorous criticism of Islam. He cannot conceive the priority he has always given to Jewish-Christian friendship as shutting the door in the face of other religions. Nevertheless, his critical look at Ellul's unilateralism does not exclude the expression of boundless gratitude.

Born in Lyon, Bernard Rordorf was minister, teacher and then chaplain in secondary schools, co-director of the Center Protestant d'Études (Protestant Center of Studies), and finally Professor of Systematic Theology at the Faculty of Protestant Theology, University of Geneva (Switzerland) from 1994 to 2005. He has always sought out means of articulation between Christian faith and society, education and adults, and the diffusion of Protestant culture. Ellul was an absolute reference and inspiration in all of this. Bernard Rordorf took great motivation from him especially regarding not only his search for a method of theological research deeply rooted in Scriptures, but also particularly his efforts to build a theology of creation and of eschatology. In the *Center Protestant d'Études* journal, in which he publishes articles and book reviews, he contributes toward popularizing the thought of Ellul (Rordorf 1974, 1985; Ellul 1985d, e). Bernard Rordorf proposes an Ellulian reading of the Book of Genesis able to emphasise its implications for life today, particularly regarding our relationship with the animal world (Rordorf 1992, 2005, 2007).

Jean Baubérot is a famous historian and sociologist of Protestantism and *laïcité*. He simultaneously studied theology at the *Institut Protestant de Théologie* (Protestant Institute of Theology) of Paris, and history at the *École Pratique des Hautes Études* (Practical School of Higher Studies). He has been particularly interested in the impact of Ellulian theology on French Protestantism during the 1960s (Baubérot 1983). His research most notably concerns the Ellul book entitled: *Fausse présence au monde moderne* (Ellul 1963). This book greatly helped the young theologians of this generation, after the Algerian War and before 1968, to structure their own thinking whilst also radically criticizing modern society and the theologies of the world. They conceive themselves as actual revolutionaries, in the Ellulian meaning of the word. But these young theologians, who Jean Baubérot himself belonged to, refused to affiliate behind any single master, and wanted to multiply the springs of inspiration, and to always remain critical toward the thinkings of Ellul himself. Jean Baubérot has moved away from Ellul, distancing himself on topics such as Israel, Islam and the Palestinian conflict (Baubérot 1970). The polemics between the two men became heated, when Ellul accused Jean Baubérot of Anti-Semitism (Ellul 1977). The relation of Jean Baubérot to Jacques Ellul is fundamentally ambivalent: he recognizes his boundless intellectual debt toward his thinkings, while simultaneously wanting to integrate this debt into his own framework. Hence, he would never be able to identify himself as an Ellulian.

In turn, Jean-François Hérouard is a militant ecologist and the deputy mayor of the city of Cognac (South-West of France), with a sustainable planning portfolio. A student of Ellul at the Faculty of Law and at the *Institut d'Études Politiques* in Bordeaux, he frequented the *Fédé*, the national group of protestant students, where he took on some responsibilities. He later organized training courses for students at the Faculty of Protestant Theology in Paris. Paradoxically, his engagements in the municipal politics of Cognac are strongly inspirited by the example of Ellul: according to him, the Ellulian criticism of politics concerns the nation and big cities such as Bordeaux, but not small towns such as Cognac (20,000 inhabitants), where the municipal crew can master problems and orientate them toward a certain ecological and social welfare. Jean-François Hérouard tries to apply in politics the evangelical ethic principles he received from Ellul, and particularly in terms of the topics of resistance against *hubris* (Hérouard 2011).

Jean-François Zorn is the first Frenchman to devote his Master's thesis, presented at the Montpellier *Institut Protestant de Théologie* (Protestant Theological School) in 1971, to the work of Ellul (Zorn 1971). This work represents a detailed analysis of Ellul's theology, for which the author has a great deal of sympathy. Going against the grain of intellectual fashion, the thesis went unnoticed, seemingly struck by the same ostracism that affected Ellul's own work. Although Jean-François Zorn later specialized in fields such as missiology, which Ellul never explored, he continues to acknowledge his debt to Ellul. For Zorn, Ellul was less a mentor than a stimulus who enabled him to strike out in his own direction.

Michel Rodes is a Protestant geographer and a militant ecologist from Orthez (South-West France). When he was a student at the University of Bordeaux, he frequented some of the courses held by Ellul, but above all the biblical studies that the professor hosted at home. He became a friend of the family, and a spiritual son of Ellul, who baptized him when 23 years old. He went on to assume responsibilities in a local Church in Orthez, study theology, become a member of the Barthian review *Foi et Vie* staff in 1977, and join the *Bible et Création* (Bible and Creation)[11] network of Protestant ecologists. He regularly writes in the Protestant press in defense of environmental causes (Rodes 2010). He was town councillor in Orthez (11,000 inhabitants) from 1989 to 2008, and deputy mayor from 2001 holding portfolios for urbanism and the environment. This experience of municipal politics and the difficulties encountered led Michel Rodes to agree with Ellul and his critique of "political illusion" (Ellul 2004b). For his spiritual, theological and sociological legacy, Michel Rodes can only express a vast gratitude toward Ellul and his thinkings.

Olivier Abel is best known in France as a philosopher and an ethicist, specializing in the works of Paul Ricœur, that other great intellectual figure of twentieth-century French Protestantism, whose asymmetrical relationship to Ellul is manifest in his affirmation of the importance of institutions, politics, psychoanalysis and hermeneutics. Nevertheless, Olivier Abel considers himself a student of Ellul's, owing as much to him as to Ricœur: the "No" Ellulian and the "Yes" Ricœurian contributed in equal parts to the construction of his own dialectical thought. For there can be no "Yes" without a "No," a truth which each author implicitly acknowledged: Ricœur, the ethicist, protests against all that which is opposed to "aiming at the good life with and for others in just institutions"; and Jacques Ellul, the Barthian, emphasizes that God says "Yes" to humankind even as God says "No"

[11] See http://blog.bibleetcreation.com

to the works of men and women. Indeed, it seems as if the dissymmetry between Jacques Ellul and Paul Ricœur could be expressed as the subordination, in the former, of "Yes" to "No," and in the latter as an inverse subordination of "No" to "Yes." Another indication of the Ellulian influence on the intellectual and spiritual journey of Olivier Abel is the extensive televised interview that he did with Ellul for the program "Présence protestante,"[12] later published in book form (Lavignotte 2004; Abel 2012), both of which contributed significantly to the dissemination of Ellulian thinking throughout French Protestantism.

Within contemporary French Protestantism, Sylvain Dujancourt is without a doubt one of those who best knew Ellul, of whom he was both a parishioner and a student. He is also one of the foremost specialists on Ellul's writings on law, having devoted both his Master's (Dujancourt 1989) and his Ph.D. (Dujancourt 1996) theses to the question. Sylvain Dujancourt was minister in Amnéville, and then at Saint Paul's Reformed Church in Strasbourg until 2008. He also served as editor of the journal *Foi et Vie* from 1997 to 2004. Today, he is president of the International Jacques Ellul Society and member of the editorial board of the *Cahiers Jacques Ellul*. In addition, he is the author of numerous articles. His talents as a disseminator of Ellulian ideas within Protestantism and as a respected theologian in Ellulian circles make him an ideal interface between Protestants and the network of intellectual heirs to the Bordeaux thinker.

Antoine Nouis is assuredly one of the most famous French Protestant ministers of the twenty-first century. He has authored a number of bestselling books on theology, liturgy, catechism, and Bible commentaries for the general public (Nouis 2002, 2003, 2006, 2007, 2010, 2011). He writes a column for the weekly *Réforme* (and became the editorial director of this paper on January 1st 2012), is a member of the National Council of the *Fédération Protestante de France* (Protestant Federation of France), and plays an active role in propagating Protestant culture in French society. Although references to Ellul in his work are often more implicit than explicit, they are abundant. Antoine Nouis openly acknowledges the profound influence of Ellul on his own spiritual and intellectual journey.

Jean-Sébastien Ingrand, a young minister of some 40 years of age, devoted his Master's and DEA[13] thesis to the work of Ellul. The discovery of Ellul not only shaped Ingrand's thought but also influenced his lifestyle choices, determining his involvement in the degrowth movement. As director of the Protestant Multimedia Library in Strasbourg he has contributed to spreading Ellulian ideas, most notably by creating a monthly study group.

Finally, Stéphane Lavignotte, pastor and coordinator of the Maison Verte, the Parish Mission fellowship in the 18th arrondissement of Paris, after a career as a journalist, now finds inspiration for his activism in the work of Ellul and the

[12] Cf. Claude Vajda, *Jacques Ellul: Portrait/entretien: sans arme ni armure* (entretien avec Olivier Abel), France 2, 1992, 60 mn.

[13] "*Diplôme d'études approfondies,*" formerly an intermediary degree between the Master's and the PhD, the DEA has now been abandoned by the French system of higher education. Translator's note.

tradition of social Christianity. A degrowth militant and member of the Green Party, he is one of a new generation of activists for whom Ellul plays a key role in their spiritual quest and social involvement (Lavignotte 2004, 2012).

3 Conclusion

As these portraits show, the impact of Jacques Ellul owes more to personal affinity than to any mass social phenomenon. Quantitative analysis of references to Ellul would almost certainly lead to the conclusion that reception of his thought is modest in proportion, even within French Protestantism, despite a small flurry of interest over these past years. On the other hand, Ellul has made an indelible mark on the spiritual journeys and the intellectual itineraries of many individuals far beyond the few influential figures emblematic of theological and ecclesiastical debates that we have chosen as illustrations in this article. But is this paradox not in keeping with Jacques Ellul's way of thinking? He who, faithful to the Kierkegaardian matrix, spoke to each individual as a unique person irreducible to another, in order to lead her or him to make free, responsible existential decisions?

Acknowledgements Translated from French by Jason Dean.

References

Abel, Olivier. 2012. *Dialogues avec Paul Ricoeur, Jacques Ellul, Jean Carbonnier, Pierre Chaunu. Quatre passions protestantes pour le possible*. Geneva: Labor et Fides.

Baubérot, Jean. 1970. *Le tort d'exister. Des juifs aux Palestiniens*. Saint Médard en Jalles: Ducros éditeur.

Baubérot, Jean. 1983. *Le pouvoir de contester. Contestations politico-religieuses autour de "mai 68" et le document "Église et pouvoirs"*. Geneva: Labor et Fides.

Chastenet, Patrick. 1994. *Entretiens avec Jacques Ellul*. Paris: La Table Ronde.

Dujancourt, Sylvain. 1989. Introduction à la pensée juridique de Jacques Ellul. MA diss., Faculté de Théologie protestante de Strasbourg, Strasbourg.

Dujancourt, Sylvain. 1996. Étude théologique et éthique du droit dans la société technicienne. PhD diss., Faculté de Théologie Protestante de Strasbourg, Strasbourg.

Ellul, Jacques. 1963. *Fausse présence au monde moderne*. Paris: Les Bergers et les Mages.

Ellul, Jacques. 1969. *Autopsie de la révolution*. Paris: Calmann-Lévy.

Ellul, Jacques. 1972. Rapport de la Commission de réforme des études de théologie. In *LXVe Synode national, Nancy – Pont à Mousson*, 29–30 avril et 1er mai 1972, ed. Église Réformée de France, 23 + 30–48 + 206–245. Paris: Église Réformée de France.

Ellul, Jacques. 1973. *Éthique de la liberté*. Geneva: Labor et Fides.

Ellul, Jacques. 1977. Impressions d'Israël. *Foi et Vie* 76e année/4:1–72.

Ellul, Jacques. 1979. Preface to *Écoute, Kierkegaard. Essai sur la communication de la parole*, Nelly Viallaneix, II. Paris: Le Cerf.

Ellul, Jacques. 1983. En parlera-t-on dans les Synodes? Une proposition scandaleuse de Jacques Ellul: Supprimons l'institution! *Réforme* 2005: 12.

Ellul, Jacques. 1984. *Les combats de la liberté*. Paris/Geneva: Le Centurion/Labor et Fides.

Ellul, Jacques. 1985a. Pour des États Généraux du protestantisme. *BIP* 968: 9–10.

Ellul, Jacques. 1985b. Jacques Ellul propose des États Généraux du protestantisme (entretien avec Jean-Pierre Manigne). *Le Christianisme au XXe siècle* 19: 3.

Ellul, Jacques. 1985c. La démocratie dans l'Église: Pour des États-Généraux du protestantisme. *Réforme* 2113: 16.

Ellul, Jacques. 1985d. Karl Barth et nous. *Bulletin du Center Protestant d'Études* 37th year/4–5:5–12.

Ellul, Jacques. 1985e. Sur la dialectique. *Bulletin du Center Protestant d'Études* 37e année/8: 7–26.

Ellul, Jacques. 1986a. Les États Généraux du protestantisme font couler de l'encre: Jacques Ellul ne partage pas l'avis de Monique Ranson sur la rencontre de Fontevraud. *BIP* 1036: 5.

Ellul, Jacques. 1986b. Karl Barth et nous. *Réforme* 2143: 7.

Ellul, Jacques. 2004a. *L'espérance oubliée*. Paris: La Table Ronde.

Ellul, Jacques. 2004b. *L'illusion politique*. Paris: La Table Ronde.

Ellul, Jacques. 2006. *L'idéologie marxiste chrétienne. Que fait-on de l'Évangile?* Paris: La Table Ronde.

Ellul, Jacques. 2007. Présence au monde moderne. Problèmes de la civilisation postchrétienne. In *Le défi et le nouveau. Œuvres théologiques*, ed. Ellul Jacques, 53–111. Paris: La Table Ronde.

Ellul, Jacques, and Yves Charrier. 1971. *Jeunesse délinquante. Des blousons noirs aux hippies. Une expérience en province*. Paris: Mercure de France.

Guivarch, Olivier. 2008. Livre: Jacques Ellul. *Évangile et liberté* 223: 22.

Hérouard, Jean-François. 2011. Pour une écologie évangélique. *Réforme* 3413: 3.

Lavignotte, Stéphane. 2004. *Jacques Ellul. Actualité d'un briseur d'idoles. Cinquante ans de chroniques dans "Réforme"*. Paris: hors série de Réforme.

Lavignotte, Stéphane. 2010. Le christianisme social est relancé. *Évangile et Liberté* 244: 4–5.

Lavignotte, Stéphane. 2012. *Jacques Ellul. L'espérance d'abord*. Lyon: Éditions Olivétan.

Lavignotte, Stéphane, and Michel Vaquin. 2010. *Disputatio*: La décroissance, voie d'avenir ou impasse économique? *Réforme* 3375: 10–11.

Leplay, Michel. 2002. *La racine qui te porte. L'histoire mouvementée de la lecture chrétienne de la Bible juive*. Poliez-le-Grand: Éditions du Moulin.

Leplay, Michel. 2006. *Les Églises protestantes et les Juifs face à l'antisémitisme au vingtième siècle*. Lyon: Éditions Olivétan.

Moreau, Christian. 2011. La gauche et l'affaire DSK. *Réforme* 3420: 16.

Nouis, Antoine. 2002. *L'aujourd'hui de la Création: Lecture actualisée du récit des commencements*. Lyon: Réveil Publications.

Nouis, Antoine. 2003. *L'aujourd'hui de l'Évangile. Lecture actualisée de l'évangile de Marc*. Lyon: Réveil Publications.

Nouis, Antoine. 2006. *L'aujourd'hui du salut: Lecture actualisée de l'Épître aux Romains*. Lyon: Éditions Olivétan.

Nouis, Antoine. 2007. *L'aujourd'hui de la loi: Lecture actualisée des dix commandements et du sermon sur la montagne*. Lyon: Éditions Olivétan.

Nouis, Antoine. 2010. *Lettre à mon gendre agnostique pour lui expliquer la foi chrétienne*. Geneva: Labor et Fides.

Nouis, Antoine. 2011. *Un catéchisme protestant*. Lyon: Éditions Olivétan.

Rodes, Michel. 2010. Écologie et amalgames. *Réforme* 3383: 16.

Rognon, Frédéric. 2011. L'affaire DSK: un non-événement. *Réforme* 3418: 5.

Rognon, Frédéric. 2012. *Générations Ellul. Soixante héritiers de la pensée de Jacques Ellul*. Geneva: Labor et Fides.

Rordorf, Bernard. 1974. Recension de Jacques Ellul, Contre les violents. *Bulletin du Center Protestant d'Études* 26th annual 5–6: 12–19.

Rordorf, Bernard. 1985. Avant-propos. *Bulletin du Center Protestant d'Études* 37th year/8: 3–6.

Rordorf, Bernard. 1992. *Tu ne te feras pas d'image. Prolégomènes à une théologie de l'amour de Dieu*. Paris: Les Éditions du Cerf.

Rordorf, Bernard. 2005. *Liberté de parole. Esquisses théologiques*. Geneva: Labor et Fides.

Rordorf, Bernard. 2007. L'arbre de la connaissance du bien et du mal. *Cadmos: Revue Culturelle et Scientifique* 11.

Vahanian, Gabriel. 1961. *The death of God. The culture of our post-Christian era*. New York: George Braziller.

Vahanian, Gabriel. 1976. *Dieu et l'utopie: L'Église et la technique*. Paris: Le Cerf.

Vahanian, Gabriel. 1983. Espérer, faute de foi? In *Religion, société et politique: Mélanges en hommage à Jacques Ellul*, ed. Étienne Dravasa, Claude Émeri, and Jean-Louis Seurin, 153–167. Paris: PUF.

Vahanian, Gabriel. 1992. *L'utopie chrétienne*. Paris: Desclée de Brouwer.

Vahanian, Gabriel. 2008. *Praise of the Secular*. Charlottesville/London: University of Virginia Press.

Zorn, Jean-François. 1971. Deux ex machina (le Dieu machine). Présentation et interprétation de la pensée sociologique de Monsieur Jacques Ellul. MA diss., Faculté de Théologie Protestante de Montpellier, Montpellier.

Chapter 14
Radically Religious: Ecumenical Roots of the Critique of Technological Society

Jennifer Karns Alexander

Jacques Ellul formulated his influential critique of technological society in the decade following the Second World War, as one of a group of theologians and church people concerned about technology and social justice in war-torn Europe. They are a group I will call the Technology and Social Justice Movement. Their work was sponsored by the World Council of Churches in Geneva, and Ellul was its most recognizable speaker. Ellul visualized a society founded neither on Marxist nor capitalist terms, by radically rejecting the concepts of planning inherent in both. This paper analyzes the speech that brought Ellul to international attention, at the first assembly of the World Council of Churches in Amsterdam in 1948, and draws on the correspondence and papers of Ellul held in the Geneva archives of the World Council of Churches. Ellul's contributions required him to merge what he would later distinguish as his theological and sociological approaches. I argue that Ellul's Amsterdam contributions illustrate how theologically grounded and truly radical his critique of technological society was.

Amsterdam illuminates the theological understory of Ellul's criticism, because he counseled the Christian churches to oppose large administrative structures and to resist the momentum of technological developments, and to instead hold themselves accountable before God and neighbor in an immediate and local way rather than by system and plan. Ellul's criticism centered not on particular technologies; internal combustion engines did not in themselves concern him, nor did the use of fertilizers or electric lights and appliances. What concerned Ellul was the system that modern technological developments required, a system whose goal was to be efficient, and a system which, as he would later famously write, "required all things human to be analyzed and integrated into orderly and manageable systems"

J.K. Alexander (✉)
Department of Mechanical Engineering, University of Minnesota, Minneapolis, MN 55455, USA
e-mail: jalexand@me.umn.edu

H.M. Jerónimo et al. (eds.), *Jacques Ellul and the Technological Society in the 21st Century*, Philosophy of Engineering and Technology 13,
DOI 10.1007/978-94-007-6658-7_14, © Springer Science+Business Media Dordrecht 2013

(Ellul 1964: 20–21, 74, 427). Historians have yet to discuss the connection between technology and theology, between technology and the sustained and foundational intellectual effort to understand not the social or cultural out-workings of religious belief, but the effort to understand the divine itself, and humankind's relation to the divine. We cannot understand Ellul without coupling technology and theology, nor can we understand the movement of which he was a part. What distinguished the Technology and Social Justice Movement from the practice of what many called the social gospel was precisely the theological commitment of its members. One could practice social gospel, caring for the poor and aged, feeding the hungry, marching for justice, while describing Jesus as friend or brother or teacher. Those in the Technology and Social Justice Movement spoke a more rigorous creed: "Jesus Christ as God and Savior."

We can see this in the work of one of the planning commissions for the Amsterdam Assembly of which Ellul was a member. Commission III was charged with examining social disorder and identified the key issue as the rise of a technical society.[1] But the commission proposed no solution, only a problematic: that Christians should work toward a responsible society, and that what mattered wasn't achieving it – that was something God would have to do – but acting responsibly in opposition to the forces of a totalizing, technologizing world.

I will begin by surveying the ecumenical background of the World Council of Churches and its Amsterdam meeting, and the social disorder still manifest in the wake of the Second World War. Then I will turn to the substance of the Amsterdam Assembly, examine the official report of Commission III, Ellul's address to the full assembly, and his work behind the scenes in preparing the report and the articles collected in a planning volume distributed in advance to assembly delegates. I will conclude with some remarks about the importance of Ellul's theological grounding for considering the future of his work.

1 Ecumenism and the Amsterdam Assembly, 1948

In preparing for Amsterdam, Ellul worked with fellow Christians from a wide variety of denominations and confessions. The World Council of Churches was officially formed in Amsterdam in 1948, the culmination of a generation of work to overcome the divisions between different Christian traditions. The Amsterdam Assembly was the World Council of Churches' first; the correspondent from *Time Magazine* called it "the greatest Protestant gathering since the Reformation." The World Council itself grew out of two international movements with their roots in the early twentieth century, one, the Faith and Order Movement, dedicated toward bridging – "healing" was the term

[1]The Ecumenical Council had formed commissions to prepare for the Amsterdam Assembly: I, The Universal Church in God's Design; II, The Church's Witness to God's Design; III, The Church and the Disorder of Society; and IV, The Church and the International Disorder.

often used – the theological divisions that kept the various Christian denominations apart; the other, the Life and Work Movement, dedicated to the application of Christian ethics to the social problems of modern life.

Shared interests and backgrounds brought together members of the Technology and Social Justice Movement: serious theological interests, often supplemented by rigorous theological training, and backgrounds in activism at the intersection of technology and society, most often concerned with issues of work, industrial or in craft practice, both within and outside of Europe. This combination of interests and activities had motivated many people for more than a generation before Commission III started meeting in preparation for Amsterdam. What Commission III contributed was an analysis that brought together theology, technology, and social needs: a description of contemporary technological organization that placed it in contrast to a responsible society ordered to encourage the freedom, justice, and dignity of all peoples as the children of God (Report of Commission III, "The Church and the Disorder of Society," 1948: 189–197).

The Amsterdam meeting was firmly ecumenical. The official delegate list numbered nearly 500 and represented were not only the European nations and the United States, Canada, and Australia, but also Indonesia, India, Mexico, Siam, Egypt, Iceland, Brazil, Ethiopia, the West Indies, Philippines, and China. Delegates and observers came from a multitude of Christian confessions; at least 30 denominations were represented. With the exception of the American Polish Catholic Church, no representatives of Roman Catholicism attended although they had been invited. Several hundred journalists registered with the press office, and several thousand people attended as accredited observers. Combined sessions met in Amsterdam's Concertgebouw – attendees remembered with pleasure the sessions of hymn-singing accompanied by its great organ – but with a capacity of only 2,200 it was small and cramped when the hero Martin Niemöller or the theological superstar Reinhard Niebuhr spoke. The theme of the conference was "Man's Disorder and God's Design."

Observers reported a splendidly organized event: the pageantry stately and magnificent, all the ministers in their robes, the discussions friendly and courteous. People remarked on a sharp exchange between John Foster Dulles, representing the Federal Council of the Churches of Christ in America, and Josef Hromadka, professor of theology at the John Hus Theological Faculty in Prague and member of the Evangelical Church of Czech Brethren; Hromadka objected to Dulles characterizing communism as ungodly, and retorted that at least communism was feeding his people. Hromadka's Czechoslovakia was at that moment transitioning into communism following a recent socialist coup.

Matters proceeded in a calm and orderly fashion, until the report of Commission III, of which Ellul was a part, on the Church and the Disorder of Society. Loud dissent greeted its report, mainly from Americans who did not like its treatment of capitalism. Others objected that the document contained no practical directives, and one observer concluded that the document's treatment had been "ill advised" and "unworthy of the occasion" (Gaines 1966: 291). Controversial was the report's advice that Christian churches "reject the ideologies of both communism and *laissez faire* capitalism" (the *laissez faire* had been added during debate). Some were disappointed that the report did not tell the churches what to do; by condemning both economic and social poles, the document had instead created confusion and was no help in assessing "the practical potentialities of democratic free enterprise." Anyone who has read much Ellul will recognize this type of criticism; the other commissions had taken up specific questions and returned to the Assembly with concrete plans of action.[2]

What Commission III did offer was a problem: it posed the problem of technological society. Preparing for another WCC Assembly several years later, C. L. Patijn, member of Commission III, jurist, lay member of the Netherlands Reformed Church, and Counsellor in the Ministry of Economic Affairs in the Hague, recalled what had been unique about the Amsterdam report. For at least a generation the ecumenical movement had emphasized the dangers of economic life for the individual soul, but at Amsterdam Commission III decided on a different approach:

> Now for the first time… an attempt was made to give cohesive appraisal of our technical civilization in all its complexity and intractability. Society no longer appeared under a moral aspect only… [out of their study] appeared a picture of society – of a technical civilization of things, of forces and powers which man has called into being, but which dominate him as a second nature… in many respects far more hostile… than original nature itself (Gaines 1966: 640).

2 Ellul and Commission III at Amsterdam

The work of Commission III to the Amsterdam Assembly, and its report, must be considered within the context of the mess Europe was in after the war. Ellul and his compatriots in Commission III hoped that a new society might indeed be born from the collapse they daily saw. Those traveling to the Amsterdam Assembly had encountered an apocalyptic topography; and that apocalypse was as much a part of the context of the Assembly as was its ecumenical background. Delegates passed through cities still in rubble, camps remained filled with semi-starved refugees and displaced persons, lines between east and west were hardening, much of the productive capacity of the land had been destroyed. A description of the apocalyptic landscape underscores the social disorder Commission III was tasked with assessing, and also highlights the potential that Ellul and his colleagues saw for remaking society.[3]

[2] Commission I of the Assembly took up the need for church unity in a period of human disruption; Commission II considered how the church might put the Gospel to work at the social level; Commission IV examined the church's approach to international affairs, its relationship with national governments, for example, and with international bodies like the United Nations or the International Labor Office. They all made concrete recommendations: Commission I, that the Unified Church recognize and continue to discuss divisive issues; from Commission II came a three-pronged strategy of missions and evangelization; and Commission IV urged that domestic and international action be taken to encourage observance of Human Rights and Fundamental Freedoms.

[3] The relentless first chapter of Tony Judt's *Postwar* is a brilliant description of this apocalyptic topography (see Judt 2005: 13–40).

In light of the disintegration of European society that was daily apparent to participants at Amsterdam, the catastrophism that many detected in the work of Commission III, and in Ellul's work most especially, does not seem odd. Large organizations were indeed moving, both to the east and to the west. Governmental structures that would have given people the opportunity to shape their own societies had collapsed, and the church was one of the remaining communities of possible influence. In lectures on the Bible delivered at the Ecumenical Institute in 1946, in preparation for the Assembly, Ellul had described the troubling malaise he hoped to address, in sympathy with those who were "weighed down by anxiety about what the great ones of the earth are plotting, by the fear which penetrates our world, by the fear they feel for a terrible phantom which they cannot even name" (Menninger 1975: 244).

We now turn to Ellul's specific contributions to the work of Commission III. The commission's official report and Ellul's address at the Amsterdam Assembly demonstrate that his concerns over technical society were shared by many within the group. The commission's report and papers were circulated among some three-dozen reviewers in preparation for the assembly, and many of them were compiled in a preparatory volume distributed in advance to delegates at Amsterdam. It contained records of reviews, correspondence with authors, and minutes of planning meetings constituting a round-table on the question of the technical society, and the Church's role within it. Reviewers sent their comments to the Geneva office of the Ecumenical Council (which became the WCC in 1948), where they were translated and duplicated, German versions mailed to German readers, French to French, English to English.[4] In short, the official report and the papers went through extensive peer review.

That technology should be the focus of contributions on social disorder had been decided early in the planning for the Amsterdam meeting, long before Ellul became involved. J.H. Oldham, one of the founders of the World Council of Churches and instigator of Commission III, had been interested in technical issues since his days as a missionary, especially regarding how technology affected labor and industry (Oldham 1950).[5] Oldham invited Ellul to join because of Ellul's deep knowledge of the Christian scriptures and his ability to interpret modern society through their lens, something Oldham had witnessed when Ellul gave a series of lectures on the Bible in the modern age, at the Ecumenical Council at the Chateau de Bossey outside Geneva in 1946.

3 Report of Commission III

The commitment of Commission III to issues of social justice as seen through a theological lens was demonstrated in its report. The official report went through three drafts before it was presented for discussion at the Amsterdam meeting; it was

[4]Comments in languages other than these three were noted and filed, some commenters wrote on Dutch, for example, but not translated and distributed.

[5]It was this volume that Ellul objected to in the opening pages of *The Technological Society* (1964: 3). In the 1950s Oldham would go on to author an influential volume on *Work in Modern Society*.

further amended in a series of discussions that week, and the final version presented to the whole assembly and published in the omnibus volume *The Amsterdam Assembly Series* (Report of Commission III, "The Church and the Disorder of Society," 1948: 189–197). The official report contained five Commissions: the Disorder of Society, Economic and Political Organization, the Responsible Society, Communism and Capitalism, and the Social Function of the Church. "The world to-day is experiencing a social crisis of unparalleled proportions," the report began, locating its deepest root in people's refusal to admit the primacy of their responsibility to God, primary over any earthly loyalty. The report affirmed the dignity and freedom of the children of God and also the depth of evil in human nature, both of which modern society continued to underestimate. The Christian Church must approach social disorder through "faith in the Lordship of Jesus Christ"; the faith that enabled them to see the sins that had corrupted human society also assured them of ultimate victory, through Christ, over sin and death. The report was thus unabashedly Christian – ecumenical but not interfaith.

For our purposes, the most interesting part of the report was its description of the roots of the crisis of disorder. Two factors were responsible: first, "vast concentrations of power," economic power under capitalism, economic *and* political power under communism, and, the domination of society by technics. This was a domination that would only increase through technics own increasing momentum. Huge political and economic organizations had diminished peoples' personal social responsibility and had attenuated their capacity for personal action. Technics had undermined connections to "family, neighborhood and craft." The report defended technology as having "relieved men and women of much drudgery and poverty," and noted that large areas of the world would benefit from further machine production, and suggested that new techniques of communication may actually aid human fellowship. This defense of technics highlighted an on-going concern of Commission III members that they not speak too negatively of it and thus alienate their audience.

Having identified the roots of the disorder, the report went on to envision the characteristics of a society in which people could live fuller personal lives. Vitally important would be a "rich variety of smaller forms of community, in local government, within industrial organizations, including trade unions, [and] the development of public corporations and through voluntary associations." Echoing a specific contribution by Ellul, the report affirmed the primacy of human beings over technical and economic production: "Man is not made for production, but production for man." Economic and political power had to be distributed widely, and people had to be able to control, criticize, and change the authorities under which they lived and worked. Christians should recognize the appeal of communism, for in evoking equality and brotherhood it had much in common with Christian teachings. Most controversial was the report's description of the conflict between Christianity and capitalism. Capitalism subordinates human need to the economic advantage of those in power, it produces serious inequity, it encourages materialism in its emphasis on money as a measure of success, and it subjects people to social disasters like mass unemployment. The Church was called to renew itself through faith and obedience to Christ, to recognize its failures (especially in having sanctified racial distinctions in many parts of the world), and to seek to overcome social barriers to bring people together into diverse worshipping communities. The report offered no solution to the disorder it had identified, other than obedience to God.

The measured prose of the published report superseded the heated language and pointed criticisms of the drafts, where the crisis of unparalleled proportions was spelled out and the deep and shared commitment by commission members to social justice made explicit. The drafts described the crisis as occurring at the very foundations of communal life, where personal and neighborly connections were shaken and individual people rendered available for re-integration into organizations of control and regimentation, for which Nazi Germany was the model. A natural harmony had not developed out of the scientific and technical discoveries of recent centuries, and no one any longer offered society a common, acceptable goal. Instead, priests and legislators had been told to get out of the way of the "chariot of progress," which had resulted not in social harmony but in social injustice, and had condemned many to lives of poverty and drudgery. Humans had not mastered nature, but had instead made people slaves to the industrial-technological machine. "Social justice is subverted, when daily bread, and opportunities for work, leisure, education, creative art are not the common right of all, but the privilege of those who are members of controlling power groups, or enjoy their favor" (Assembly Commission III on "The Church and the Disorder of Society," Report on Commission III, First unofficial draft 1948). Social justice was, indeed, "God's will for men in history: The ultimate basis for Christian action in the world is faith that God wills an ordered community of justice and freedom and fraternity among men" (Assembly Commission III on "The Church and the Disorder of Society," Report on Commission III, Complementary Statement 1948).

Significant in this report is its citing of organizational and technological momentum as the roots of crisis, and its call for human freedom and responsibility before both God and neighbor. These are theological statements, the second (responsibility before God and neighbor) most obviously. But so is the first: the momentum undermines human ability to act responsibly before God, and is thus a social evil.

4 Ellul's Amsterdam Address, August 25, 1948

Ellul's address followed Kathleen Bliss's rousing condemnation of the scientific and technological society, which, as she put it, threatened society with destruction not from without but from within: at the apex of the most advanced technological activity was the atomic bomb, and its attendant horrors (Bliss 1948). Bliss was editor of the *Christian News-letter*, a collaborator of Oldham's, and a mainstay in the English ecumenical movement. She and Ellul had been chosen to introduce the work of Commission III to the full assembly at Amsterdam. It must be noted that Ellul did not present his speech himself; his second son Simon, just 6 years old, had recently been killed in an accident and Ellul asked to be excused from the meeting, having entered upon a time of great sadness and fatigue. His address was instead presented by another member of Commission III. The address was delivered in French; English and German translations were distributed to delegates. The address was not published.

Bliss's address concentrated more on the technical society than would Ellul's and she offered an apologia he would not: that people must not condemn technology out of hand, but must recognize all the benefits it had brought to people's lives. This apology also appeared in the Commission's report, as discussed above. "Which of us here," she asked, "does not owe the life or health of some loved person to the skill of a doctor or the scientific and technical resources of laboratory and hospital?" Yet the omnipresence of technics worried Bliss, who contributed a chapter to the Amsterdam preparation volume on personal relationships in a technical society. Family and marriage relationships, household layout and the growing importance of office work all showed the influence of technical developments, but so did the lengthening human life span and society's ability to educate its young rather than putting them right to work. For good or for ill, technical society was spreading even to the pastoral regions of Africa, India, and China, where leaders sought to import the industries that had brought prosperity to the now devastated West. Bliss spoke of technical society but, in contrast to Ellul, offered no criticism of what might properly be called the technical. Her criticisms centered instead on the destruction of cherished ways of life and the church's need to choose wisely among new possibilities. Bliss spoke more forcefully than Ellul of the concerns that had motivated Commission III.

Ellul's speech was circumspect, and it illustrates the commitment to social justice he shared by other members of the commission. He sounded themes that will be familiar to those who know his later work: the need to describe situations in a concrete rather than abstract or theoretical manner, the lordship of Jesus Christ that carries hope amidst despair, and the need for the church to repent of its own disorder. The substance of disorder was made clear in Ellul's address more so than in any other contribution of Commission III. He described it as suffering of all kinds, as people lacking shelter, lacking food, as concentration camps still to be closed, and displaced persons ranging across the continent, prices rising, and dictators seizing power.

Ellul's address noted the two main characteristics of the disorder identified by the commission: the rupture of personal relations and the depersonalization brought about by the growth of great organizations (he put it more colorfully: people "cease[d] to be human beings in order to become undifferentiated cells of social groups or robots" [Ellul 1948]), and the technical aspect of disorder, recognizing technics as a cause of the disorder and that techniques of organization continued as part of the problem. The Church was called to help preserve a livable order, by which Ellul meant an order that allowed people to live fully human lives while remaining receptive to the Gospel. The Church was certainly in the world but was not required to use the world's methods of action, and Ellul argued that when the Church modeled itself on the world it interfered with its own mission. What was needed was what he called "a kind of inventory of the present world" (Ellul 1948), a listing of the world's values, what it considered its truths, its institutions and means of action. From this inventory the Church should select the elements capable of maintaining true human life and preserve them. The Church needed also to help nourish the new civilization to come, and on this point Ellul was explicit: this meant planning, coordinating actions, creating new institutions.

Striking about this address are two things: Ellul's acquiescence in advocating the use of systematic planning, and his restraint in condemning and describing the technical society. He wrote the address entirely himself. It was not vetted or reviewed. He sent a hand-written manuscript to Geneva, and it was transcribed and translated without editing. It is tempting to think that here we have the unvarnished Ellul. But these two factors suggest that he was aware of his role as a representative of the commission and spoke as such, rather than using the forum to mount his own critique. For Ellul had already formulated a position against strategic planning – he was very precise about what sort of planning people could do – and he had already concluded, from his study of the early chapters of Genesis, that technology, technique, was not a part of God's created order but was developed after humans had descended into sin.

5 Ellul Behind the Scenes: Chateau de Bossey, June 1947

Behind the scenes were just as important, for there Ellul revealed the truly radical and unique elements of his thought, in meetings, conversations, and papers circulated in preparation for the Amsterdam gathering. Behind the scenes he revealed his coupling of, and rejection of, techniques of planning and administration as part of the very technological order he criticized, and his unwillingness to consider technology as a part of the order of creation, maintaining instead that it was part of the fallen world of human construction. He made these points most clearly in a week-long meeting in June of 1947, when Commission III discussed papers for the preparatory volume to be distributed in advance to Amsterdam delegates.

The setting was intimate: the Chateau de Bossey, east of Geneva, briefly owned by Madame de Stael, which the World Council of Churches had rented for some years and which was shortly to be purchased for the WCC by the Rockefellers. It is a lovely setting, crossed by mountainside hiking trails and with views down to the lake. Those attending lodged in adjacent rooms and took meals together. What was preserved in minutes and correspondence is merely part of the on-going conversation of the meeting, but it is enough to demonstrate how different Ellul's views were even from his deeply religious, deeply Christian, compatriots.

Ellul had remained largely silent during the formal discussions, perhaps because they were held in English, which was the chair's language but one in which Ellul had never felt comfortable. Ellul suggested more attention be given to the foundations of the justice that the commission discussed in its draft report, and he stated strongly that, when issues of governance and the state came up, the form of the state did not matter – capitalist, socialist, communist – what mattered were the methods it used, increasingly totalitarian methods in a time of mass society. He had reported briefly on organizations in France that drew people who were actually practicing technological occupations into conversation. He had rejected an argument by Paul Tillich (in a circulated paper; Tillich did not attend the meeting) that technical reason was a forerunner of planning reason: planning reason was instead "really the technical reason applied to other subjects," Ellul argued. Ellul could have developed any of these remarks at great length. We hear in them early soundings of his later and deeper critique. Here, in 1947, they demonstrate how unique was his contribution, even within a gathering of Bible-reading, praying, committed Christians.

Ellul did take part in a sustained argument on Wednesday morning, June 25, at a discussion of J. H. Oldham's draft paper, "Man, Machine, and Society." C. L. Patijn, lay member of the Netherlands Reformed Church and Counsellor in the Ministry of Economic Affairs in the Hague, opened by observing that Oldham's paper was too negative toward technology, and asked whether it was "right for the Church to say this so one-sidedly?" Technics were part of nature, and we knew nature's dangers – "why should we then condemn the machine? (Wednesday morning, 9:30 June 25th; Minutes of Meeting of Commission III 1948: 1)." Oldham had himself worried about depicting technology negatively, insisting that the group "must not condemn the machine itself, since it was only the wrong use which man made of it which had created the present evils." He wished to "affirm the machine itself, distinct from the misuse made of it by man"("The meeting continued on June 24th, at 8:15 p.m.," Amsterdam Commission III, 1948: 2–3).

Ellul took this as an opening. He thought the Commission had agreed that machines were God's gift, but Patijn's criticism suggested the question was still open. Ellul summarized the view: "In Genesis we are told that man was created to have dominion over the world, and technics is a means to this end. Technics is also a part of nature, and therefore indiscutable." But before acceding to this view Ellul argued they should decide whether the machine was indeed a gift from God "or whether it is a catastrophe." We, now, can predict Ellul's answer: it was a catastrophe. He rejected descriptions of the machine as neutral: "I do not agree, for I do not believe that there is anything neutral in the world. The world is the domain of Satan who is under the lordship of Jesus Christ. Since man has fallen, he can only use the machine for evil purposes. Man is not free to choose between good and evil in the use of the machine." Ellul rejected the idea that people could master their machines, and indeed he said he was surprised to hear the argument. He used the image of floodwaters: "I see men under the domination of technics, like a torrent which we cannot stop ("Wednesday morning, 9:30 June 25th," Minutes of Meeting of Commission III, 1948: 1–4)."

Ellul did not go unchallenged. M. M. Thomas, of the World Student Christian Federation and former secretary of the Mar Thoma Syria Church of Malabar, suggested Ellul thought that scarcity was "more blessed than plenty." Ellul responded that the Church shouldn't try to direct technical activities, but that it should render a verdict on values. The question to be asked was whether India, for example, would be better off without the machine, but "'better' according to what criterion?" Emil Brunner, professor of systematic theology at Zurich, said Ellul had his theology all wrong; things that were part of the order of creation should be considered good, and that justice must be based on the order of creation. Ellul responded by quoting Proverbs: "'Oh God, preserve me from poverty and also from riches.' If we are poor, we cannot find our way to God; if we are rich, we deny God." Then he described his theological point: values could not be based on the order of creation, because creation had been perverted by the human introduction of sin, by the Fall – the human fall, through human disobedience, from communion with God.[6] The Church had to preach the Gospel, and thus to preserve the world to hear the Gospel. Another participant objected that the Bible enjoined humans to be fruitful and to multiply, and that technical progress had enabled people to escape the scenario envisioned by Malthus. Ellul did not reply, but held his peace for the rest of the meeting.

People familiar with Ellul's work will hear nothing unusual in the meetings at Bossey in June of 1947. But the point is not to find something new in Ellul's early thought. It is, instead, to show that much of Ellul's critique of technique was shared by a wide range of people within the ecumenical movement, and that the ecumenical movement provided Ellul with both an early forum for his technological critique and an early venue for testing his ideas. In this exchange we do have something like the unvarnished Ellul, speaking his own point of view, not as a representative of Commission III but as a member of an inquiring group.

The meetings do demonstrate how radical was Ellul's thought, even among his fellow believers. The positions Ellul took in this meeting, that planning was itself an extension of technical reason and that technology was not part of the order of creation but reflected human activities after the Fall and was thus neither neutral nor good, appeared neither in the commission's report, although Ellul helped to draft it and engaged in discussions of it, nor in Ellul's own address to the Amsterdam Assembly. The exchange illustrates the limits of the overlap between Ellul's thinking and that of other members of Commission III. It illustrates fundamental differences on the theology behind varying condemnations of the technological society. Ellul was concerned with technical reason and technical activity at its most basic: human attempts to build a human world. Oldham and others in the commission were concerned about machines and industrial production and their effects on human life and community. Ellul had been concerned about production, too, but as a new sort of god under which society was organized. Oldham, Bliss, and others thought the problems lay in specific forms of industrial organization, not in the very goal of measurable productivity itself.

6 Conclusion

We are tasked with evaluating the importance of Ellul's work for the twenty-first century. If Ellul's seminal analysis of technology is to remain relevant, it must be made clear how it differed from the many other critiques of technological society

[6]This was a point that Ellul would develop in the coming years in his studies of the early chapters of Genesis (Ellul 1960, 2010); see also Vanderburg (2010).

circulating at the same time. One place to look for that difference is in Ellul's early and quite public contributions to the first Assembly of the World Council of Churches, held in Amsterdam in 1948. I have emphasized two differences between Ellul's thought and that of others mounting similar critiques: his rejection of planning as itself a technique of human power, and his refusal to consider technology as part of the order of creation and thereby susceptible to good use as well as abuse. These differences mean that Ellul's critique applied not only to industrial society, as did Oldham's and Bliss's, for example, but that it will also continue to offer insights into the increasingly common technological underpinnings even of cultures with vastly different productive and religious traditions.

Ellul's contributions to the work of Commission III of the World Council of Church's Amsterdam Assembly demonstrates how thoroughly grounded in theology was Ellul's critique of technical society. Ellul's critique was informed by long years of Bible reading and study, and by debates with prominent theologians and leaders in the Christian churches. By 1948 he had already worked out many of its themes. But Commission III did not embrace Ellul's emerging critique, and within a few years Ellul had withdrawn from the group.

References

Ellul, Jacques. 1960. La technique et les premiers chapitres de la Genèse. *Foi et Vie* 59: 97–113.

Ellul, Jacques. 1964. *The technological society*. Trans. John Wilkinson. New York: Vintage Books.

Ellul, Jacques. 2010. *On freedom, love, and power*. Toronto: University of Toronto Press.

Gaines, Edward P. 1966. *The world council of churches: A study of its background and history*. Peterborough: Richard R. Smith Noone House.

Judt, Tony. 2005. *Postwar: A history of Europe since 1945*. New York: Penguin Press.

Menninger, David C. 1975. Jacques Ellul: A tempered profile. *Review of Politics* 37: 236–246.

Oldham, O.H. 1950. *Work in modern society*. London/New York: SCM Press/WCC & Morehouse-Gorham.

Vanderburg, William H. 2010. How the science versus religion debate has missed the point of Genesis 1 and 2: Jacques Ellul (1912–1994). *Bulletin of Science Technology Society* 30: 430–445.

Archival Sources

Assembly Commission III on "The Church and the Disorder of Society," Report on Commission III (First unofficial draft), A2 Study Organization d. Study Organization 1946–1948, Box 24.090, Archives of the World Council of Churches, Geneva, Switzerland.

Assembly Commission III on "The Church and the Disorder of Society," Report on Commission III (Complementary Statement), A2 Study Organization d. Study Organization 1946–1948, Box 24.090, Archives of the World Council of Churches, Geneva, Switzerland.

Bliss, Kathleen. 1948. "'The Church and the Disorder of Society', An Address at the First Assembly, World Council of Churches, August 24, 1948, Amsterdam," WCC Assembly Amsterdam 1948, Box 31.003, file 9.1. Archives of the World Council of Churches, Geneva, Switzerland.

Ellul, Jacques. 1948. Amsterdam statement. World Council of Churches, WCC Assembly, Amsterdam, 1948. Box 31.003, file 9.2. Archives of the World Council of Churches, Geneva, Switzerland.

Report of Commission III "The Church and the Disorder of Society". n.d. (meeting held in 1948), Received by the Assembly and commended to the churches for their serious consideration and appropriate action, *Man's disorder and god's design: An omnibus volume of The Amsterdam Assembly Series,* New York: Harper & Brothers, V. III.

"The meeting continued on June 24th, at 8:15 p.m.," pp. 2–3, Amsterdam Commission III, "The Church and the Disorder of Society," I, Study Department Amsterdam Assembly, Commissions II and III, Witness (II) and Society (I) vom 1947 bis 1948, Box 24.057, Archives of the World Council of Churches, Geneva, Switzerland.

"Wednesday morning, 9:30 June 25th, 1948. Minutes of Meeting of Commission III" Amsterdam Commission III, "The Church and the Disorder of Society," I, Study Department Amsterdam Assembly, Commissions II and III, Witness (II) and Society (I) vom 1947 bis 1948, Box 24.057. Archives of the World Council of Churches, Geneva, Switzerland.

Chapter 15
Truth, Reality and the Ten Commandments: Not for Theology Alone

Virginia W. Landgraf

The purpose of this paper (much of which appeared in a different form in Landgraf 2003) is to construct a reading of the Ten Commandments in terms of Jacques Ellul's thought in order to facilitate the understanding of his thought as a whole, both theological and sociological. For the most part, this is not an exposition of Ellul's statements about each commandment but a construction based on his statements about what it means to be human and what keeps societies from being destroyed from within by vicious circles of power. In the space allotted I must paint in broad strokes for entire chapters could be written about each heading here. My thesis is that Ellul's conception of the space within which life is possible can be delineated by the Ten Commandments, paraphrased in terms relating to his understanding of the orders of truth and reality, and that the outcome of obedience to the commandments so paraphrased is that reality is kept open to truth. A side result is that Ellul's arguments about the autonomy of technique depend on his belief that fallen human beings have an innate desire to possess reality.

1 Truth and Reality

A distinction between *truth* and *reality* is fundamental to Ellul's thought. The distinction is most fully articulated in Ellul 1981 (9–42 [English translation (henceforth ET) Ellul 1985c: 5–37]), but it appears throughout his work (e.g., Ellul 1997a: 168–180; Ellul 1975c: 239 [ET Ellul 1970: 165]). Some of the ramifications of this distinction raise important questions, such as the relationship between love

V.W. Landgraf (✉)
American Theological Library Association, 300 S. Wacker Drive Suite #2100,
Chicago, IL 60606-6701, USA
e-mail: kaencat@sbcglobal.net

H.M. Jerónimo et al. (eds.), *Jacques Ellul and the Technological Society in the 21st Century*, Philosophy of Engineering and Technology 13,
DOI 10.1007/978-94-007-6658-7_15, © Springer Science+Business Media Dordrecht 2013

and power, which cannot be resolved by this paper. However, these are intramural disputes compared to the task of keeping reality open to truth.

The distinction between truth and reality is not merely of different content but of two different "orders," each with its own characteristic mode of transmission and logic and inviting a certain attitude toward the world. Truth has to do with human beings' ultimate destiny and values that have the same ultimate seriousness; reality includes that which can be seen and measured in space, expressed in terms of visualizable abstractions, or appears as given in various spheres of life (such as political or economic realities). Truth is communicated by the word of a committed person and requires time to absorb and debate; reality can be transmitted impersonally as abstract data and grasped immediately. Truth invites a posture of waiting and listening toward the other; reality invites an attitude of grasping at and holding power over these objects which seem so ready to be manipulated. Attending to truth goes along with allowing for love and freedom in relationships; grasping at reality brooks no opposition, but ends up with the human being enslaved by the reality which he or she supposedly possesses, yet is completely dependent upon.

This distinction being made, Ellul does not think that all answers to the question of truth are equally legitimate. He believes that the one Truth worthy of the title is the God who created the world, led the Israelites out of Egypt, became incarnate in Jesus Christ, and will bring the world to its final consummation. Before the Fall, according to Ellul, Truth wholly penetrated reality; after the Fall, we think we can arrogate reality to ourselves, but are cruelly deceived. Between the Fall and the final consummation, only in the incarnation of Jesus Christ has truth completely penetrated reality (Ellul 1981: 89–92, 280 [ET Ellul 1985c: 79–82, 253]).

1.1 "You Shall Not Kill": Nonviolence vs. the Realm of Necessity

It would not be too far off to equate the "realm of necessity" in Ellul's thought with how reality holds together after the Fall. Before the Fall, Ellul thinks, creatures spontaneously obeyed the will of God; after the Fall, various forms of law–physical, moral, civil – are needed to keep creatures from destroying each other out of the naked will to power. These laws neither have the spontaneity of God's love toward creatures (or of creatures toward God before the Fall), nor do they bring the life which only relationship with God can bring. Creatures experience them as external constraints or fates (Ellul 1964b: 52–55 [ET Ellul 1969b: 59–62]). One seems to be faced with a choice between a pre-ordained destiny within these laws and an earlier death if one violates them. Yet Ellul can also speak of creatures as meeting death because they follow the laws of necessity: because certain kinds of necessity prescribe killing (such as predators' biological needs or civil laws requiring war or capital punishment), and because they contain a law of reciprocity: violence reproduces violence (Ellul 1972a: 119–120, 124–126, 185 [ET Ellul 1969c: 95–97, 101–103, 146]).

"You shall not kill" (Ex. 20:13, Deut. 5:17) is therefore a commandment rising beyond these necessities. Ellul cites this commandment as marking the boundary between the human and the animals in both sociological and theological works (Ellul 1975d: 78 [ET Ellul 1978: 65]; Ellul 1972a: 185 [ET Ellul 1969c: 146]). This paper will not attempt to resolve the exegetical problem of what kinds of killing the Hebrew *ratsach* refers to but focus on how Ellul's expansive definition of violence fits in with keeping reality open to truth. Ellul includes under "violence" not only killing but any trauma, physical or psychic, which inhibits another's future development as a self-directing agent (Ellul 1972a: 122–123 [ET Ellul 1969c: 97–100]). The posture of nonviolence is thus equivalent to the posture of attending to truth – listening, waiting, love, and freedom; the posture of violence is equivalent to the posture of manipulating reality, and especially of reducing a human being capable of responding to Truth into a mere object of reality to be possessed. One can therefore paraphrase the commandment against killing in Ellulian terms as "You shall not deny another person's existence or prevent his or her future development as a self-directing agent."

Here one may question whether Ellul's stark distinction between a love that waits and a power that crushes has accounted for all the alternatives, both in dealing with human beings and in interacting with the rest of creation. What about grabbing a child's hand away from the fire? Are there some ways of structuring civil law codes that are more likely to promote the development of citizens as self-directing agents than others? What part do words or reciprocal, truth-telling communication play in lovingly tending to animals or in artistic creation that is respectful of the materials it works on – neither of which seem reducible to power that crushes? The place of what one might call "benevolent coercion" is a significant problem in Ellul's thought. However, debates over its presence or absence do not concern whether reality should be open to truth, but whether Ellul has fully accounted for all the ways in which God is active in the world after the Fall.

1.2 The Prologue of the Ten Commandments vs. Non-committal Sociology

In both Exodus and Deuteronomy, the Ten Commandments are prefaced with God's self-identification: "I am the Lord your God, who brought you out of the land of Egypt, out of the house of slavery" (Ex. 20:2; Deut. 5:6). This prologue plays a decisive part in the difference between the readings of the commandments according to Ellul's theological and sociological writings. Theologically, it may be paraphrased in Ellulian terms as, "The only Truth worthy of trust and obedience is *YHWH*, who created the world, brought the Israelites out of Egypt, became incarnate in Jesus Christ, took upon God's self the world's sins in Christ's crucifixion, and will lead the world to its final consummation. This God gives human beings freedom in history, brings judgment upon human works which do not conform to God's will, destroys powers opposed to God's will in the last judgment, and will have communion with each human being in the New Jerusalem." But when reading the commandments according to Ellul's sociological work, the prologue is empty: "No Truth, no One Guarantor is promising that you will be able to fulfill any of these preconditions for individual or societal resilience."

Theologically, the prologue to the commandments is the key to how a law can become the antidote to the laws of necessity. Ellul believes that it is only in relationship with this God that the commandment functions this way. Detached from this God, it is possible for these commandments to take their place in the framework of necessities and for a legalistic religion to focus on reality instead of truth (Ellul 1984c: 84–89 [ET Ellul 1986: 69–73]). The prologue to the commandments is an integral part of all that follow, even those that do not refer to God. In a statement that functions as a general theological anthropology in his commentary on Ecclesiastes, Ellul says that fearing God and obeying this God's commandments are the sum of what it means to be human (Eccl. 12:13; Ellul 1987b: 277 [ET Ellul 1990: 291]). This is not a servile fear but a relationship with a God who waits for human beings to respond and sometimes gives them what they want even though it was not God's original design. A paradigmatic example in Ellul's thought is the city; although the construction of cities epitomizes human beings' rebellion against God by trying to build an environment entirely of their own construction, God does not return them to the garden in the final consummation but takes up human history in the new Jerusalem (Ellul 1975c: 30 [ET Ellul 1970: 26–27]; Ellul 1975a: 236–238 [ET Ellul 1977a: 221–224]; Ellul 1987a: 284–287 [ET Ellul 1989: 218–221]).

Ellul follows Karl Barth in treating the commandments as promises: that those in relation to God will not have to kill, commit adultery, steal, and so forth (Ellul 1981: 68–69; ET Ellul 1985c: 60–61). The commandments delineate the space within which life is possible. The societal forces and psychological drives that urge us on to violence, adultery, and various lusts for unequivocal reality are not aids to liberation from a restrictive code of behavior but accomplices in the web of necessities that ends in death. As long as one is determined by these lusts one is not free. Just think of what someone may confess after being overwhelmed by personal desire, social pressure, or the apparent requirements of the situation to break one of these commandments and follow through on the impulse: "I couldn't help it; I just *had* to do it." It may be a contingent action at the individual level, but it fits in with forces that function in the fallen world as inexorable necessities. Much of Ellul's theological work is meant to function as testimony that the God beyond these necessities has broken into history decisively in Jesus Christ, can be addressed by human beings now, and promises a future beyond these death-dealing forces.

Ellul offers no such guarantee in his sociological work. According to his theological method, knowledge of God comes by revelation, not neutral knowledge (Ellul 1981: 54–56; ET Ellul 1985c: 48–51). In his sociological writings he claims to be explicating the workings of societies in ways that should, in principle, be understandable by all. Testimonies to the living God who may intervene in these workings – even though Ellul believes that God may be silent (Ellul 1972b: 112–114 [ET Ellul 1972c: 110–112]) – belong to theology, not sociology. It is not surprising, then, that in his sociological work there is a preponderance of descriptions of vicious circles, such as how the technical phenomenon drives politics (Ellul 1977b: 64–65 [ET Ellul 1967: 42–43]), yet, in turn, the state gives a sanction to the technical phenomenon (Ellul 1954b: 207 [ET Ellul 1964a: 228]). Power lustful of reality squeezes out truth and reinforces power. (Ellul 1947: 704–709 [ET Ellul 1997b: 56–61])

However, a less prominent side of Ellul's sociological work provides the key to understanding how a "prologue-less" version of the Ten Commandments has relevance beyond his theological work. In a few places he shows signs of what might be called, for lack of a better term, a positive social philosophy or philosophical anthropology: a vision of sustainable life for selves and societies that is not expressed in theological terms. He draws on entropy theory to show that a society needs to have tensions between various groups and opinions within it in order to be resilient enough to meet the challenges it faces; without the capacity to receive new information it will eventually die (Ellul 1977b: 292–293 [ET Ellul 1967: 209–210]). He calls for contemplation as part of a the kind of revolution required to counter the obsession with following the course of history and being caught up in the momentum toward an ever more powerful and violent state (Ellul 1969a: 334–335 [ET Ellul 1971: 285–286]). He thinks that personality formation is a prerequisite for citizens' responsible participation in public debate and that reason, self-control, and respect for the other are prerequisites for the development of a coherent personality not tossed about by every whim and circumstance (Ellul 1977b: 325–331 [ET Ellul 1967: 232–238]; Ellul 1975d: 50–61 [ET Ellul 1978: 38–49]). The common denominator of all these prerequisites is openness to dialogue with the other and within various constituents of the self or one's experience, such as reason and passions, continuity and change, and so forth. In terms of the distinction between truth and reality, this vision of sustainable life requires openness to truth. The various commandments in the Decalogue can be paraphrased in terms of different aspects of this openness. In Ellul's sociological work, the commandments so paraphrased still delineate the space within which life is possible, but Ellul does not testify in this side of his writings to a Guarantor that this space will be inhabited.

1.3 Keeping Selves and Societies Open to Truth: Trust, Thought, Imagination, and Desire

The commandments against having gods other than YHWH (Ex. 20:3; Deut. 5:7), making graven images and worshiping them (Ex. 20:4–5; Deut. 5:8–9), and coveting or lusting after anything that belongs to someone else (Ex. 20:17; Deut. 5:21) can be easily paraphrased in terms of the distinction between truth and reality, where the only distinction between the theological and sociological versions would be the substitution of "truth" for "God" in the latter. "You shall not trust in reality, even invisible spiritual realities that manifest themselves with wordless power, as if it or they were God/truth." "You shall not reduce truth to reality, or bow down to any reduction of truth to reality or worship it." "You shall not grasp at reality apart from

God/truth." It is somewhat arbitrary to label the first as having to do with what one trusts, the second as concerned with what one worships (and how one conceives it), and the last as having to do what one desires (apart from any formal acts of worship), because trust, identification of what one worships, and desire are not easily separated out. But between these three commandments, a whole spectrum of temptations is mapped out. The living God has spiritual rivals, such as the spirits of financial or political power, which one can set before oneself as a "god" without having a conscious realization that one is worshiping something false or a specific visualization of their effects. One may intend to worship YHWH but box the self-revealing God into a construction of our own determination. One may desire to take possession of reality apart from God or truth in a variety of ways: conceiving this reality as somehow unrelated to God and neutral; seeing it as being possibly manipulated by us; or wanting to possess it physically. Again, as in the commandment against killing, these readings of the commandments in terms of Ellul's thought are expansive, giving a thick description to the boundary between freedom and unfreedom.

The tie between covetousness and Ellul's description of how we encounter reality is very important in Ellul's thought. Ellul thinks that covetousness is the sum of all other sins and that one's attitude toward this commandment expresses one's basic attitude toward God and the world (Ellul 1981: 112 [ET Ellul 1985c: 101]; Ellul 1975b: 122). He also thinks that fallen human beings have in our minds an image of us possessing reality that causes us to want to grasp it for ourselves: in short, to covet. He even describes this image in terms from the story of eating the forbidden fruit in Genesis 3: seeing that the fruit appears attractive, then being drawn to take and eat it (Ellul 1981: 12–17 [ET Ellul 1985c: 7–12]). Thus Ellul would think it difficult for our minds to remain in a neutral position about reality, once we imagined it as unrelated to God (as I speculated in the paragraph above). The belief that fallen human beings have this image in our minds ends up functioning as a doctrine of original sin in Ellul's work, and even in sociological works where theological terms or arguments are absent. In the last section of this paper I will show how it is fundamental to understanding the logic of technical autonomy, even in Ellul's sociological works.

1.4 Keeping Language Open to Bearing Truth

Two of the commandments specifically refer to language: the one against improper use of God's name (Ex. 20:7; Deut. 5:11) and the one against bearing false witness against one's neighbor (Ex. 20:16; Deut. 5:20). These can be considered to be about resistance to propaganda and other uses of language that betray the word's distinctive function as truth-bearer. Ellulian paraphrases of these commandments could be, "You shall not use language in an empty way so as to rob it of its capacity to witness to God/bear truth," and, "You shall not report about another person in such a way as to either reduce the truth of their being to reality or distort the realities about them." Ellul thinks that, contrary to popular belief, propaganda is about changing people's behavior, not their beliefs (Ellul 1962: 36–38 [ET Ellul 1973: 25–27]). Even when couched in the phrases of high-sounding values, it functions in the realm of observable results. It is not concerned with communication between committed persons but with using mechanized methods to shape and move masses. Its use of words as power tools may crowd out quieter, more patient uses of words that encourage witnessing to and listening for truth (Ellul 1981: 38–39 [ET Ellul 1985c: 32–33]). Such reductive uses of language thus fuel closed systems in the realm of reality, acting in opposition to healthy tensions within societies and healthy personality formation.

Reportage about other people that reduces the truth about their being to an aspect of their reality is similarly unhealthful for selves and societies. An example is the use of labels or characteristics – whether accurate or inaccurate – as accusations that enable someone's beliefs and perspectives to be dismissed, or, conversely, as guarantees that their views are automatically acceptable. Communist, fascist, Republican, Democrat, Muslim, Christian, gay, straight, celibate, married, single, alien, citizen, legal, illegal – all such aspects of reality can be used to praise, blame, exalt, or castigate above and beyond (and, alas, instead of) nuanced and historically informed discussions of public policy, theological truth, sexual ethics, or immigration law where such distinctions would have relevance. Even if the label is an accurate representation of the person's opinions and self-identification, a person is always more than his or her political views, religious affiliation, theological beliefs, ethnic background, familial situation, immigration status, or any other aspect of reality.

One may ask whether false witness in the realm of reality is as important as false witness in the realm of truth. Ellul does believe that the difference between the orders is such that a wrong answer to the question of truth, which he calls falsehood, is of a different order from an incorrect answer to questions of reality, which he calls inaccuracy (Ellul 1981: 37 [ET Ellul 1985c: 32]). It is possible that someone's testimony about his or her neighbor is about questions of reality, such as where the neighbor was at 6 in the morning on Friday. But the mere fact that a witness is needed implies that this is not (yet) impersonal information that can be communicated immediately. And the answer may have bearing on the person's character and relationships. Furthermore, some of the poles that need to be balanced in Ellul's vision of healthy individuals and societies, such as continuity and change, include non-human realities. Therefore, false witness about realities cannot be dismissed as irrelevant to the task of keeping realities open to truth.

1.5 Time, Work, and… Property?

The commandments to keep the Sabbath and against stealing can be linked by a common thread: work and human material sustenance. Ex. 20:8–10 and Deut. 5:12–15 both enjoin the Israelites to keep the Sabbath holy to YHWH and for all human and animal members of their households to abstain from work on that day. The two versions give different reasons for keeping the Sabbath: that God rested on the seventh day (Exodus) and that God brought them out of Egypt, the house of bondage (Deuteronomy). Either way, the commandment to keep the Sabbath desacralizes work. Writing during the Cold War, Ellul sees the competing pro-capitalist and Marxist ideologies of the day as both exalting human work and its results at the expense of values beyond material prosperity: one more way in which reality is valued over truth (Ellul 1954a: 20–25 [ET Ellul 1984b: 20–23]). Sociologically, he subscribes to a view of economic history whereby poverty has increased with the availability of paid labor, so he believes that these ideologies cannot even deliver on their promise to free people from poverty (Ellul 1982: 159–160; Ellul 1980: 4 [ET Ellul 1985b: 45]). Theologically, Ellul sees no reason to give work ultimate status, because he believes that human beings will be separated from our works at the last judgment, with only some of our works being found worthy of entrance into the new Jerusalem – and we have no way of knowing which ones those are (Ellul 1987a: 284 [ET Ellul 1989: 217–218]). We should therefore do the work which falls to us without obsessing over its results, casting it upon the waters (cf. Eccl. 11:1) and trusting that it may be of help in some way (Ellul 1987b: 101 [ET Ellul 1990: 102]). The commandment to keep the Sabbath puts work in its place, reminding us that discernment of truth requires time and cannot be reached by our productivity. Labor-saving devices that supposedly buy us time are of no help in this respect if we simply fill our leisure hours with ever more distracting realities. Ellul characterizes those who are numbed by technical work and leisure styles as a new proletariat (Ellul 1982: 210). Thus in Ellulian terms, the commandment to keep the Sabbath may be expressed as, "You shall leave time for God/truth in your life, for your work can never be the truth of your life."

Ellul's critique of Cold War absolutisms also affects how one may interpret the commandment against stealing (Ex. 20:15; Deut. 5:19) in terms of his thought. At first, the commandment seems straightforward: stealing means following through on covetousness, taking possession of something one wants that belongs to another. How can Ellul, who sees covetousness as the basic sin, not also oppose stealing? On the other hand, Ellul recognizes that rules about property and stealing have varied across societies. He sees the absolutization of private or state property as a distortion of legitimate ends of property: to provide for the security of the individual's living space (Ellul 1984a: 322 [ET Ellul 1976: 481]) and to be used for the meeting of society's material challenges, especially the needs of the poor (Ellul 1954a: 60–61 [ET Ellul 1984b: 50–51]). Since these needs, concerning human beings as they do, cannot be exhaustively defined in terms of reality (cf. the injunction against reducing the truth about someone else's being to reality), institutional absolutization of a particular way of assigning property fails to keep reality open to truth. Moreover, Ellul sees that ideologies about property often fit into systems of self-justification: people assume that institutional changes will solve society's problems and rest content when they have acted toward that aim, failing to recognize their own complicity in evil or the continuing need to be open to truth (Ellul 1954a: 13–19 [ET Ellul 1984b: 15–19]). Keeping these ideas in mind, one may paraphrase the commandment against stealing as, "You shall not arrogate property to yourself as an expression of covetousness, nor shall your institutions absolutize private, collective, or state property so as to obscure property's true end: to provide the requisite continuity for meeting the material challenges of individuals and societies, especially the needs of the poor."

1.6 Honoring Face-to-Face Relationships and Resisting Objectification of the Body

The remaining two commandments, those enjoining the honoring of one's parents (Ex. 20:12; Deut. 5:16) and against adultery (Ex. 20:14; Deut. 5:18) have to do with maintaining family relationships. Ellul has a positive view of the family throughout his work, seeing it as functioning to keep people together in enduring face-to-face relationships (Ellul 1987a: 101–111 [ET Ellul 1989: 73–82]). Honoring one's father and mother can be seen as the positive complement of the commandment not to do violence: do not merely avoid damaging others, but listen to those who brought you forth – those whom you did not choose – that you may learn to listen for truth. Have patience with and care for your aging parents and ill spouse in their physical weakness, that you may learn that love is more important than physical results. The commandment to honor one's parents can thus be paraphrased as, "Honor your mother and your father, that the conditions for genuine dialogue between persons capable of responding to God/truth can be sustained." "[C]onditions for genuine dialogue…" is an appropriate substitution for "that your days may be long and that it may go well with you in the land that the Lord your God is giving you" because it underscores the connection between dialogue and social and personal resilience in Ellul's thought.

With respect to the commandment against adultery, an expansive paraphrase that discourages fornication is in keeping with Ellul's thought: "You shall not engage in sex outside of marriage, because it inhibits the development of relationships where people can see the other as a person worthy of dialogue and not an object." Ellul sees the pledged word in marriage as significant (Ellul 1985a: 196–197). The spouses promise to care for each other regardless of changing realities (again, keeping reality open to truth). It is not to be supposed that unmarried couples or even people in adulterous relationships are prevented from having conversations about truth or from all instances of loving or giving behavior by the irregularity of their sexual union, nor that the promises of marriage guarantee a non-abusive relationship, in which the spouses' bodies (or, indeed, their minds and whole beings) are not treated as objects to gratify immediate and possessive desires. Rather, it is that in marriage – by contrast with institutions such as money and the state, which Ellul sees as based on impersonal abstractions and reinforcing vicious circles of power squeezing out truth – a promise is made that goes beyond existing circumstances or sexual desire.

Table 15.1 The ten commandments in terms of Ellul's theological and sociological thought

Commandment	Interpretation in terms of Ellul's thought and the distinction between truth and reality	
Prologue: "I am the God who brought you out of Egypt, out of the house of bondage"	Theological interpretation: The only Truth worthy of trust and obedience is *YHWH*, who created the world, brought the Israelites out of Egypt, became incarnate in Jesus Christ, took upon God's self the world's sins in Christ's crucifixion, and will lead the world to its final consummation. In relationship with this God, you will be able to follow these commandments	Sociological interpretation: No Truth, no One Guarantor is promising that you will be able to fulfill any of these preconditions for individual or societal resilience
Against other gods	You shall not trust in reality, even invisible spiritual realities that manifest themselves with wordless power, as if it or they were truth	
Against graven images	You shall not reduce truth to reality, or bow down to any reduction of truth to reality or worship it	
Against improper use of God's name	You shall not use language in an empty way so as to rob it of its capacity to witness to God/bear truth	
Honoring the Sabbath	You shall leave time for God/truth in your life, for your work can never be the truth of your life	
Honoring parents	Honor your mother and your father, that the conditions for genuine dialogue between persons capable of responding to God/truth may be sustained	
Against killing	You shall not deny another person's existence or prevent his or her future development as a self-directing agent	
Against adultery	You shall not engage in sex outside of marriage, because it inhibits the development of relationships where people can see the other as a person worthy of dialogue and not an object	
Against stealing	You shall not arrogate property to yourself as an expression of covetousness, nor shall your institutions absolutize private, collective, or state property so as to obscure property's true end: to provide the requisite continuity for meeting the material challenges of individuals and societies, especially the needs of the poor	
Against bearing false witness	You shall not report about another person in such a way as to either reduce the truth about their being to reality or distort the realities about them	
Against coveting	You shall not grasp at reality apart from God/truth	

1.7 *Schematization*

The previous interpretation of Ellul's thought in terms of the Ten Commandments has treated the commandments in terms of related clusters of ideas, not in their scriptural order. For ease of reference, the above table summarizes the sociological and theological paraphrases of the commandments in the order in which they are usually known (Table 15.1).

2 Postscript: Covetousness and the Logic of Technical Autonomy

A significant part of Ellul's logic about technical autonomy depends on his belief that mathematics yields unequivocal results. For instance, in *The Technological Society*, he says:

> There is not a choice, strictly speaking, with respect to size, between, say, 3 and 4. Four is greater than three. This does not depend on anyone. No one can change it or say the contrary or personally escape it. Decision, as regards technique, is actually of the same order. There is no choice between two technical methods. One of them imposes itself inevitably, because its results are counted, measured, seen, and indisputable (Ellul 1954b: 75, translation mine [published ET Ellul 1964a: 80]).

Yet consider the equation $x^2 - 7x + 12 = 0$. Its solutions are three and four. How will we obtain the one solution which Ellul expects? We need another statement of mathematical properties – such as Ellul's above, that four is greater than three – or a concrete application, such as the desired consistency of a batch of cookies, where the choice between three and four measures of flour can make a big difference, or a construction task, where supporting a structure on three or four beams changes the whole design. But why choose? Why not let three and four coexist, like musicians playing polyrhythms?

The introduction of deciding criteria between multiple results of a mathematical calculation requires a will to exclusiveness going beyond the first problem and its solutions. Ellul's logic makes sense if one assumes that most people seek this kind of unambiguity when encountering an impersonal, non-dialectical entity. There is nothing sacred about a single mathematical point as defining the exclusive answer; what matters is that the answer is defined, fixed, and therefore controllable. This kind of reaction toward mathematical solutions is a subset of the will to grasp reality – in theological terms, covetousness – that Ellul believes is inherent in fallen human beings. We have a lust for the unequivocal. We want answers that can be immediately possessed and controlled, and mathematical calculation seems to provide these.

Seen in this way, technical autonomy, as Ellul describes it, looks like an expression of original sin. Its consequence is the death of what is distinctively human, just as "the wages of sin is death" (Rom. 6:23). Not mathematical results but our will to control them forestalls waiting for questions from the realm of truth. We see calculative power as inviting our mastery but end up being its slave. Ellul's belief that technical autonomy arose with such virulence only in the last few centuries does not contradict the pervasiveness and universality of original sin. Where more sacral constraints existed around the use of mathematical knowledge, the tendency to grasp at reality apart from truth would have expressed itself in different ways. Such historical speculations go beyond the purview of this paper. However, the lacuna in his logic shows that Ellul's theological work, much of which can be schematized according to the Ten Commandments read in light of the distinction between truth and reality, is vital to understanding the logic of his sociological work.

3 Where Do We Go from Here?

One may ask what this paraphrase of Ellul's work in terms of the Ten Commandments has accomplished, beyond a neat schematization of seemingly disparate areas of Ellul's thought. It gives a thick description of the space where life is possible. To those accustomed to thinking of ethical commandments in terms of casuistry, it will seem at once more strict and less specific. It seems more strict, because it is against all forms of violence, not just killing, and against fornication as well as adultery. Yet many of the commandments seem less specific. When do we know when we have crossed the line from, say, being open to truth to reducing truth to reality and worshiping our reductions? Ellul would not want us to rest content with thinking that we had avoided idolatry simply by removing images from worship.

Another cluster of issues is raised by the possibility, which I have alluded to above, that Ellul's distinction between truth and reality sets up a distinction between love and power that does not account for all the ways in which God relates to the world or human agents interact with each other or other members of creation. What if we need a third term between the love that speaks and waits and the power that crushes – say, artistic creation that takes account of the materials that it is working on?

It seems that positing such a way that God acts, or human beings may act, would not affect the need to keep reality open to truth. From a theological point of view, biblical images of the God who interacts with creation by means other than words (e.g., the potter shaping clay, Jer. 18:4) coexist with the doctrine that the Word is God (e.g., John 1:1). If one posits such a third term between speaking and crushing, one may presume that its exercise will be guided by the character of God or the character of the agent. From a human point of view, Ellul's thought already includes the point of view that human personality formation and character are dependent on keeping reality open to truth. From the point of view of the doctrine of God, one would need to ask, "What is God's character?" This is by definition a question of truth. Ellul rejected a traditional doctrine of providence whereby all things were willed by God, thinking that it implied an unbiblical mechanistic determination (Ellul 1987a: 207–210 [ET Ellul 1989: 156–158]; a fuller discussion of Ellul's complex doctrine of divine action is found in Landgraf 2003: 17–38). However, one does not have to accept Ellul's doctrine of divine action to believe that God has a character and that one may ask whether purported divine actions would be in keeping with this character.

Finally, one may ask about the implications of this schematization of Ellul's thought for those who do not adhere to religions that see the Ten Commandments (including their prologue) as authoritative. On one level, the contrast between the "empty" sociological prologue and the "full" theological prologue may look like an apologetic. If you are concerned about the usurpation of the domain of truth by the domain of reality, you are urged to find (or rather, be found by) the living God. Ellul would certainly be in favor of anyone's coming into an active relationship with the living God! But because he thinks that such a relationship has to come into being by God's self-revelation, he would probably hesitate to label his thought as an apologetic. Instead, one of the virtues of this schema is that it enables one to make use of Ellul's thought without presuming to agree with him on every detail. It invites one to ask the questions, "Is this the problem?" and "In the face of this problem, do you have hope, and, if so, where?" Anyone who has experienced, within oneself or in one's observations of society, what I have called above the "lust for the unequivocal" will be able to relate to its depiction of the problem, even though Ellul's logic about technical autonomy may not be watertight. Adherents of non-biblical religions or worldviews may be invited to see how their hope gives them resources for overcoming the domination of truth by reality.

References

Ellul, Jacques. 1947. Le réalisme politique (Problèmes de civilisation III). *Foi et Vie* 45(7): 698–734.
Ellul, Jacques. 1954a. *L'homme et l'argent (Nova et vetera).* Neuchâtel: Delachaux & Niestlé.
Ellul, Jacques. 1954b. *La technique ou l'enjeu du siècle*. Paris: Librairie Armand Colin.
Ellul, Jacques. 1962. *Propagandes*. Paris: Librairie Armand Colin.
Ellul, Jacques. 1964a. *The technological society*. Trans. John Wilkinson. New York: Alfred A. Knopf.
Ellul, Jacques. 1964b. *Le vouloir et le faire: Recherches éthiques pour les chrétiens*. Geneva: Labor et Fides.
Ellul, Jacques. 1967. *The political illusion*. Trans. Konrad Kellen. New York: Alfred A. Knopf.
Ellul, Jacques. 1969a. *Autopsie de la révolution*. Paris: Calmann-Lévy.
Ellul, Jacques. 1969b. *To will & to do: An ethical research for Christians*. Trans. C. Edward Hopkin. Philadelphia: Pilgrim Press.
Ellul, Jacques. 1969c. *Violence: Reflections from a Christian perspective*. Trans. Cecelia Gaul Kings. New York: The Seabury Press.
Ellul, Jacques. 1970. *The meaning of the city*. Trans. Dennis Pardee. Grand Rapids: William B. Eerdmans.
Ellul, Jacques. 1971. *Autopsy of revolution*. Trans. Patricia Wolf. New York: Alfred A. Knopf.
Ellul, Jacques. 1972a. *Contre les violents*. Paris: Le Centurion.
Ellul, Jacques. 1972b. *Espérance oubliée*. Paris: Gallimard.
Ellul, Jacques. 1972c. *Hope in time of abandonment*. Trans. C. Edward Hopkin. New York: The Seabury Press.
Ellul, Jacques. 1973. *Propaganda: The formation of men's attitudes*. Trans. Konrad Kellen and Jean Lerner. Repr. ed. New York: Random House Vintage Books.
Ellul, Jacques. 1975a. *L'Apocalypse: architecture en mouvement*. Paris, [N.p.]: Desclée.
Ellul, Jacques. 1975b. *Éthique de la liberté*, vol. 2. Geneva: Labor et Fides.
Ellul, Jacques. 1975c. *Sans feu ni lieu: Signification biblique de la Grande Ville*. Paris: Gallimard.
Ellul, Jacques. 1975d. *Trahison de l'Occident*. Paris: Calmann-Lévy.
Ellul, Jacques. 1976. *The ethics of freedom*. Trans. Geoffrey W. Bromiley. Grand Rapids: William B. Eerdmans.
Ellul, Jacques. 1977a. *Apocalypse: The book of revelation*. Trans. George W. Schreiner. New York: The Seabury Press.
Ellul, Jacques. 1977b. *L'illusion politique*, 2nd ed. Paris: Robert Laffont.
Ellul, Jacques. 1978. *The betrayal of the west*. Trans. Matthew J. O'Connell. New York: The Seaburg Press.
Ellul, Jacques. 1980. De la Bible à l'histoire du non travail. *Foi et Vie* 79(4): 2–8.
Ellul, Jacques. 1981. *La parole humiliée*. Paris: Éditions du Seuil.
Ellul, Jacques. 1982. *Changer de révolution: l'inéluctable prolétariat*. Paris: Éditions du Seuil.
Ellul, Jacques. 1984a. *Éthique de la liberté*, Les combats de la Liberté, vol. 3. Geneva: Labor et Fides.

Ellul, Jacques. 1984b. *Money and power*. Trans. LaVonne Neff. Downers Grove: Inter-Varsity Press.
Ellul, Jacques. 1984c. *La subversion du christianisme*. Paris: Éditions du Seuil.
Ellul, Jacques. 1985a. La Bible est un livre de questions: N'y cherchons point de réponses. In *La famille contre les pouvoirs: de Louis XIV à Mitterrand*, ed. Kaltenbach Pierre-Patrick. Paris: Nouvelle Cité.
Ellul, Jacques. 1985b. From the Bible to a history of non-work. *Cross Currents* 35(1): 43–48.
Ellul, Jacques. 1985c. *The humiliation of the word*. Trans. Joyce Main Hanks. Grand Rapids: William B. Eerdmans.
Ellul, Jacques. 1986. *The subversion of Christianity*. Trans. Geoffrey W. Bromiley. Grand Rapids: William B. Eerdmans.
Ellul, Jacques. 1987a. *Ce que je crois*. Paris: Bernard Grasset.
Ellul, Jacques. 1987b. *La raison d'être: Méditation sur l'Ecclésiaste*. Paris: Éditions du Seuil.
Ellul, Jacques. 1989. *What I believe*. Trans. Geoffrey W. Bromiley. Grand Rapids: William B. Eerdmans.
Ellul, Jacques. 1990. *Reason for being: A meditation on Ecclesiastes*. Trans. Joyce Main Hanks. Grand Rapids: William B. Eerdmans.
Ellul, Jacques. 1997a. Christian faith and social reality. Chap. 7. In *Sources and trajectories: Eight early articles by Jacques Ellul that set the stage*, 166–183. Trans. and ed. Marva J. Dawn. Grand Rapids: William B. Eerdmans.
Ellul, Jacques. 1997b. Political realism. Chap. 3. In *Sources and trajectories: Eight early articles by Jacques Ellul that set the stage*, 51–91. Trans. and ed. Marva J. Dawn. Grand Rapids: William B. Eerdmans.
Landgraf, Virginia W. 2003. *Abstract power and the God of love: A critical assessment of the place of institutions in Jacques Ellul's anthropology of dialectical relationships*. PhD diss., Princeton Theological Seminary.

Chapter 16
Social Intolerability of the Christian Revelation: A Comparative Perspective on the Works of Jacques Ellul and Peter L. Berger

Andrei Ivan

During the twentieth century, the Christian Church became increasingly aware of a new social milieu and tried to make its message relevant in a progressively secularized culture. The social approach became dominant, with the result that, in the words of one critic, "we have come to see just how the terms 'social' and 'society' have so insinuated themselves that we never question the assumption that while 'religious' is problematic, the 'social' is obvious. The idea that the former should be referred to the latter appears like an innocent, genial inspiration" (Milbank 1993: 102).

Tracey Rowland describes three ways to look at culture from a Christian point of view. In the first, culture is described as neutral; in the second it is considered a *preparatio evangelii*; in the third it is portrayed as hostile to the Christian faith (Rowland 2003: 2).

Jacques Ellul and Peter L. Berger adopt the last of Rowland's three options and attempt to expose the intolerability of Christian revelation in a (post)modern culture and society. Although Ellul and Berger come from different theological orientations, they agreed regarding the relationship between social and religious attitudes in the modern world. Ellul was an active Christian in the French Reformed Church, while Berger defined himself as a "Christian, though I have not yet found the heresy into which my theological views would comfortably fit" (Berger 1969a: 10), which situated him closer to liberal Protestantism.

After a careful scrutiny of their writings, we can say that similarities between these two authors are not limited to their ideas on the subject under study. In fact they share almost the same starting point in their methodologies. Ellul, in his multidisciplinary writing, criticized the commonplaces (Goddard 2002: 120) of modern society and Berger exposed what is taken for granted by the modern consciousness. Another feature common to Ellul and Berger is the use of the

A. Ivan (✉)
Draganul Street, No. 8N, Bragadiru, Ilfov, Romania
e-mail: andrei.ivan1@gmail.com

H.M. Jerónimo et al. (eds.), *Jacques Ellul and the Technological Society in the 21st Century*, Philosophy of Engineering and Technology 13,
DOI 10.1007/978-94-007-6658-7_16,

dialectical method in their complex theories. In his writings, Ellul often had recourse to this, perhaps because of the influence of Karl Marx, but it is also present in Berger's writings, especially in *The Social Construction of Reality* (co-authored with Luckmann).

1 Peter Berger: Christian Faith and the Modern World in Mutual Antagonism

Berger's oeuvre can be divided into two main periods. In his early years, during the 1960s and 1970s, Berger was more concerned with sociology of knowledge theory, as can be seen from works such as *The Social Construction of Reality: A Treatise in the Sociology of Knowledge* (1967). This was the most important period of his work on the sociology of religion, and bore fruit in *The Sacred Canopy: Elements of a Sociological Theory of Religion* (1969b), *A Rumour of Angels: Modern Society and the Rediscovery of the Supernatural* (1969a), *The Precarious Vision: A Sociologist Looks at Social Fictions and Christian Faith* (1976), and *The Heretical Imperative: Contemporary Possibilities of Religious Affirmation* (1980). After this period he wrote extensively from a neo-Weberian perspective, discussing the links between religion and societal development in different regions of the world. Berger accounted for his freedom to make this transition without reassessing his previously held positions by commenting, "I found the sociology of knowledge paradigm of my early work very useful and have not been motivated to exchange it for another" (Berger 2001: 454).

In the first period, Berger's sociology of religion took account of the Christian faith as one that rejects all other absolute claims (Berger 1998: 782–796). But in writings from the last three decades he has been more preoccupied with the role of religion in social binding and as an economic catalyst. He has written of social and economic developments which are the "functional equivalents" of Weberian Protestantism: Pentecostalism in Latin America, Africa, and South Korea; *Opus Dei* in Spain; and Confucianism in Japan. Berger even suggests, for example, that evangelical congregations in Latin America serve as schools for democracy and for social mobility (Berger 1999: 1–18).

We will be more concerned with Berger's ideas from the first period, which resurface in some recent writings. In his texts of the early period we can distinguish between the legitimating role of religion and its revolutionary stance. Berger addressed both, but we will focus on the latter. Berger considered the "tendency of religion is to be socially functional rather than dysfunctional" (Berger 1976: 111). Bernice Martin observe that this position represents a mutation from his earlier thought presented in *The Noise of Solemn Assemblies*, "that the Christian vision is inherently athwart social convenience" (Martin 2001: 167) This contradiction between Berger's earlier and later descriptions of religion can be resolved by the antithesis between religion and faith that Berger embraced, following Karl Barth (Berger 1976: 163).

1.1 Religion as World Maintaining and World Shaking Force

For Berger society is a stage, a counterfeit reality that needs to be deconstructed. Theology and sociology share the same endeavor, aiming to dismantle this counterfeit reality and show the social drama as a comedy (Berger 1976: 204–205, 211). This can be achieved by theology in its prophetic stance. Though from different standpoints, theology and sociology both question, doubt, and even challenge the culture that "proclaims" itself every day (Berger and Luckmann 1967: 23) – the view that takes reality as granted, in other words, the "okay world" (Berger 1976: 121). In this sense religion is a world-shaking force (Berger 1969b: 100).

For Berger, one truth is that theology relativizes reality, but another is that modernity undermines religious worldviews through the pluralization of choices, so that *the heretical imperative* – the imperative to choose – is unavoidable. Religion is no longer based on social "plausibility structures," but is rather a matter of choice. Although this means that religion can no longer be taken for granted, it does empower and legitimate actors on the social stage. Religion is therefore a world-maintaining force (Berger 1969b: 100). Even though Christian religion itself is a world maintaining force, the Christian faith undermines social assumptions at their very roots, making religion no longer conservative but revolutionary (Berger 1969a: 203). In Berger's view, religious experience is a real threat to social organization; no society could survive a continuous encounter with the supernatural (Berger 1980: 49–50). For this reason the church cannot find itself at ease in its social and cultural milieu. This social intolerability of faith is sustained in Berger's view both by the opposition of the Hebrew God to the Mediterranean religions and by the opposition of the Hebrew religion itself to the prophets of the Old Testament. God called his people "out of the Egypt of deceptive social safety, but also out of the Zion of deceptive religious security" (Berger 1976: 180). Christian faith is not a revolutionary doctrine, but because it can relativize any morality, or any system of law and order, it cannot sustain any society or political system, whether conservative or liberal. Neither can it be the basis for a religiously sober way of life, this kind of religiousness functioning to prevent ecstasies (Berger 1976: 174–176). This prevention is accomplished by the institutions and traditions of the Church. In general, institutions control human behavior by defining behavioral norms and by preferring some directions over others (Berger and Luckmann 1967: 55). Tradition has two roles that are in conflict with each other. One is the ongoing mediating of religious experience through successive generations; another is the domesticating of such experience in order to incorporate it into regular and orderly social life (Berger 2004: 133). One way to achieve the latter is to adapt religious experience to the cultural background, which means nothing less than a "cognitive surrender," because the outside challenge is internalized (Berger 1969a: 36).

Having in mind the characteristics of the Christian faith, and also its need of an institution and tradition in order to survive, Berger asks what kind of institution would be able to transmit the faith in a way most faithful to its spirit. In his "precarious vision" he speaks about "weak" institutions based on fragile individual commitment and choice, which can also be full of vitality. This edifice must be built on the institutionalization of "permanent reflection," scepticism, and a rejecting of all taken-for-granted realities (Berger 1998: 782–796).

2 Jacques Ellul: The New Iconoclasm

Ellul is well known for his sociological analysis of the technological milieu, but not all who know his sociology also know or welcome his theological works. His sociological and theological thinking nevertheless complement each other, as do the sociology and theology of Berger. The link between the two strands of Ellul's work is that his theological position provides the basis for his severe criticism of society. At the heart of his theology lies the relationship between God and the world, viewed as a ruptured communion (Goddard 2002: 62). By contrast with Berger's, his theological stance places a great deal of importance on the Fall in the understanding of human existence. Ellul's sociology actually represents his personal reflection on reality, the aim of which is the deconstructing of commonplaces and the sacred (Goddard 2002: 120). Ellul does not understand the sacred in the accepted sense of the word.

In religious belief the sacred existed in the relationship of humans with nature, but through the desacralization of traditional religion, nature, and the cosmos, a new sacred emerged in the milieu of a technological society. This new "sacred" is not seen as such and, indeed, paradoxically, is based on the very forces of the earlier desacralization process, which have now imposed themselves as a new sacred by virtue of their achievements (Ellul 1975: 65–68). The unquestioned assumptions of modernity, precisely because they are unquestioned and for this very reason become a new but unrecognized sacred.

Both Berger and Ellul attack what is taken for granted, but Ellul's focus is wider than Berger's. He turns his new iconoclasm against the taken-for-granted reality of the technical, cultural, social, and political background. Ellul's iconoclasm is based on his faith in God and aimed at destroying the modern gods of utility, money, efficiency, and politics (Ellul 1975: 225). Taken-for-granted reality is based on the axioms of the secularized world, the coming of age of "modern man," and the assimilation of almost everything into the cultural (Ellul 1975: 210). The supremacy of the cultural imposes the normality of what a group accepts as good and moral.

2.1 Subversive Faith and Anti-subversive Religion

Society even tries to assign a role to the church; the church is accepted as long as she limits her purposes to increasing social functionality (Ellul 1975: 24, 28). Sometimes Christians themselves bind the role of the church to social preoccupations. Often, because of the uncriticized presuppositions of their worldview, people try to redefine Christianity according to their cultural and social surroundings. In this instance, not only does faith become assimilated by society, but even Jesus himself can be recast to conform to human needs and desires (Ellul 1975: 157, 214).

Ellul disapproves of the adaptation of Christianity to the modern age, but also of efforts to "Christianize" the state, society, institutions, and morals. For him, both these strategies represent an effort "to put an end both to the scandal that the world ought to be for faith and that faith ought to be for the world" (Ellul 1989a: 7–8). In his view, the Christian is the meeting point between the two opposing forces of the Word of God and the will of the world (Goddard 2002: 107). This will expresses itself by the social reality that remains incontestable in people's experience.

Ellul observed that Christ subverts any kind of power, but that Christians are conservative and anti-subversive, or in other words, Christian praxis is subverting the truth of Christ. This subversion takes place when the church, on the basis of a false idea of the conquering of the powers of this world by the Holy Spirit, makes use of these powers to advance a mission. But what in fact happens is that the church and its mission become penetrated by the powers, and her truth is corrupted. Another route toward the subversion of Christian faith is the reaction of the social body in self-defense, against what is foreign to itself, because "real people in any society, flesh and blood people, cannot swallow it" (Ellul 1989b: 158). Consequently, faith is domesticated by the social body as a result of its force of integrating, absorbing, and assimilating (Ellul 1989b: 21). In this way, in Ellul's view, "Christianity becomes an empty bottle that the successive cultures fill with all kinds of things" (1989b: 18).

Along with Berger, who sees in religion a legitimization authority, Ellul thinks that every power the church attempts to use in fact uses the church for its own legitimization, at the cost of that which is truly the church. In this way grace becomes "a politics of give and take," with the church maintaining itself at a price (Ellul 1989b: 126–127). The institutional Church is, for Ellul, the image of the subversion of Christianity since it cannot be organized because of the truth of Christ. If the truth of Christ subverts even the institutionalization of the Church, it is even less possible for it to be the organizing principle of a whole society.

The truth of Christ sets up a relationship of exhaustive conflict between revelation and the social body, a situation Ellul calls "the intolerability of revelation" (Ellul 1989b: 157). This resonates with Berger's idea that it is a dangerous sign when the church exists without being in a significant state of tension vis-a-vis culture (Martin 2001: 154–188). Consequently, Ellul believes it is much more natural for Christians to make a "gentlemen's agreement" with their culture, because of the intolerability of God's grace, which is visible in the rejection of the Son of God (Ellul 1989b: 172). This kind of agreement is made possible only by leaving to one side the fact that Truth has been crucified by Reality (Ellul 1973: 165).

2.2 *Why Is Revelation Socially Intolerable?*

In *Living Faith: Belief and Doubt in a Perilous World*, Ellul (1983) presents his view of the social intolerability of revelation in a thorough manner. He argues that revelation is socially intolerable because it introduces radicalism, destabilization, unpredictability, and powerlessness into human life.

The *radicalism* of revelation results from the free grace of God that wipes away any human ability and achievements, and from the demand for total submission to the will of God that is synonymous with the Good (Ellul 1983: 149–150). In Ellul's view,

> Because this radicalism is unbearable, people invariably succeed in gaining control of the contents of revelation. That is to say, they objectify it, transform a momentary illumination into a permanent establishment, a promise into law, hope into an institution, love into a series of works and charities, the Holy Spirit into a jurist, and the explosion of the Word into rituals and feasts. God's will *hic et nunc* is turned into a rigid commandment, dialogue into the catechism, symbolic offerings into a kind of purchase, death to oneself into good deeds, truth into dogmas, grace into a system of predestination, chosenness into privileges and superiority, and the gift of salvation into damnation for others. In effect this is separating the word from the one who pronounces it. People seize the contents of revelation, when the revelatory moment can never be reduced to such objectified contents, to things that have been analyzed, flattened, expurgated until they can all be satisfactorily adapted to meet religious needs (Ellul 1983: 153).

The radicalism of the Gospel is clearly shown in Jesus's affirmation, "You have heard… but I say to you" (Matthew 5), addressed to a nation that was proud to know the will of God, showing the new meanings of what the chosen people had known very well and for so long.

For Ellul revelation is not just radical but it is also *destabilizing*. "The sword of the Word severs the strongest ties of nature" (Ellul 1983: 142). Social convergences are shattered by revelation, despite the fact that religion founds and sustains communities. The process of degeneration from the heights of revelation to the platitudes of religion has been dominated by the mundane necessities of stability and security (Ellul 1983: 155).

An example of the destabilizing effect of revelation can be found in Ellul's *The Technological Society*, in which he argues that from the fourth century onward, Christianity has obliterated the Roman *technique* of organization, construction, industry and transport. From the point of view of organization, society was in complete anarchy, being "a-capitalistic" and "a-technical," based on customs and lacking technical will, or what he calls the "technical state of mind." In the later period of the Middle Ages, a more coherent society was developed through architectural and scholastic techniques. If the first was driven by a religious impulse, the latter is, for Ellul, the only entirely new technique developed in this period. The coming into being of this state of affairs under Christian auspices was due to the attitude of moral judgment according to which Christians have always tended to scrutinize all human activities (Ellul 1964: 32–38).

Unpredictability is another feature of revelation, and at the same time another reason for Ellul to describe it as socially intolerable. Revelation does not meet human concerns or human needs, but it brings surprises and the unexpected, as profoundly articulated in the invocation, "Not as I will, but as You will" (Matthew 26, Mark 14, Luke 22). Ellul thinks

> this is unbearable for the religious mind, which aims at getting hold of, or actually being, the place where the divine phenomenon occurs. It aims at regularizing and normalizing the acts of that unknown reality. Free grace is, inevitably, the only way God expresses himself, if he is God, but that won't satisfy any of our lofty sentiments, any of our needs

> for security, for models, for a fixed and calculated absolute, for ethical systems and social organizations. On the contrary, grace is the force that disturbs us and throws us off balance, puts us in front of a mirror and throws us back upon ourselves, while at the same time it liberates us – but without our having any merit or the right to deal with God as his equals (Ellul 1983: 153).

God is totally free to ask for the killing of one long-awaited son, or for the total destruction of entire populations, and yet he gave His Son for the many. Grace is the full and impenetrable expression of the freedom of God to have mercy on whom He chooses (Romans 9:15).

Grace reveals, also, the total *powerlessness* of human beings to meet God's standards by any human power or by any human means. This powerlessness of humans is absolutely intolerable, making revelation itself intolerable as well. The expression of this intolerability of powerlessness can be seen in religious efforts to acquire as much power as possible. For Ellul, powerlessness goes far beyond nonviolence, being a choice not to use power when you have access to it. Thus, powerlessness does not mean the lack of power, but the power to refuse power. This message is intolerable in a society that has the means to achieve unlimited power through science, technology, politics, or economics. In this context Christians need to adapt the means to their ends. Just as Jesus refused power in fulfilling his mission, Christians also must renounce political, economic, and technological power, refusing both the spirit of the age and the means that it employs, thus placing themselves in a very problematic situation (Ellul 1989c: 149–151).

2.3 The Standpoint of Relativizing the Relativizers

Compared to Berger's position, Ellul goes beyond observing the social intolerability of the Christian revelation. Ellul defines the role of the Christian in the contemporary world. He believes that this role can be portrayed by the answer to the question, "How can we be the question that God puts to the world?" In his view, only those actions that rely on the Wholly Other can mobilize the independence and power required to oppose the assimilating tendencies of sociological forces (Ellul 1972: 142). Only from this standpoint can everything be questioned and relativized. Christians are able to disclose what is taken for granted based on the outside point of view provided by the Holy Spirit. From this angle Ellul can criticize the phenomenon of *technique*, which he sees as exteriorizing human capacities. He observes that *la technique* is not merely a neutral tool, but one that destroys values and eradicates choice (Ellul 1989a: 58–76). It also creates and disseminates new values (Ellul 1980: 149). It wipes out the real possibility of human choice by imposing the principle of effectiveness, so that the human agent cannot really decide, but must choose the most effective means.

From the same standpoint he criticized what he calls "the Moloch of fact," including all facts, as taken-for-granted. By this he means what Berger calls institutions. Ellul proposes to eliminate the worship of any established fact, which he rejects insofar as they have become the final reason and the criterion of truth (Ellul 1989a: 26–30).

This revolutionary take on the modern world arises from the role of Christians as delineated by Ellul. Another aspect is the renunciation of all illusions and historic hopes (Ellul 1989b: 172). This includes the acceptance of one's place and of the intolerability of one's situation, caught between two cities and unable to renounce either the one or the other. Moreover, the Christian is called to bring the future into the present, but only to make the world tolerable. He is called not to diminish the opposition between the will of the world and the Word of God, but the opposition between world disorders and God's preserving order (Ellul 1989a: 35–41). The essence of his position is that the true Christian serves the world by opposing it, or, more precisely, by opposing its suicidal tendencies, its death wishes (Goddard 2002: 77).

Ellul describes the church that is able to nurture this kind of Christianity in the same vein as Berger. His description is worth quoting in its entirety.

> God's order is not organization and institution (cf. the difference between judges and kings). It is not the same in every time and place. It is not a matter of repetition and habit. On the contrary, it resides in the fact that it constantly posits something new, a new beginning. Our God is a God of beginnings. There is in him no redundancy or circularity. Thus, if his church wants to be faithful to his revelation, it will be completely mobile, fluid, renascent, bubbling, creative, inventive, adventurous, and imaginative. It will never be perennial, and can never be organized or institutionalized. If the gates of death are not going to prevail against it, this is not because it is a good, solid, well-organized fortress, but because it is alive; it is Life that is, as mobile, changing, and surprising as life. If it becomes a powerful fortified organization, it is because death has prevailed. Thus even on the humble level of the church, revelation cannot be organized or experienced socially (Ellul 1989b: 157).

In an opinion that is similar to Ellul's, Berger tells us that the history of the church is the history of the Christian experience and lives, not the history of dogma or institutions. So also will be its future (Berger and Luckmann 1967: 187). Berger believes that a church dedicated to a socially intolerable revelation will institutionalize a permanent reflection.

3 Conclusions

The role of religion in modern society will always be controversial. I have tried here to bring to light two views that are easily neglected when this theme is on the table. The ideas of Berger and Ellul can benefit both theologians and sociologists. Theologians may be too prone to accept a cultural interpretation of the faith, and sociologists too prone to accept the paradigm of religion as an element of social functionality. Both will find in Berger's and Ellul's writings rich alternative interpretations well worth exploring.

The views of the two authors presented here are mutually reinforcing, even though Berger is more sociological in his approach and Ellul more theological. They share common ground when they speak about the dialectic of religion as a legitimating/undermining social force, about religion and society trying to domesticate a faith that unsettles both religion and society, and about the need for a more fluid church.

Both authors accept that the Christian faith is socially intolerable and expose what is taken for granted in modern society. They diverge when they choose a standpoint from which to mount their critique of society. Berger is more descriptive in his sociological deconstruction of the social reality of religion, and Ellul is more prescriptive of the theological and sociological role of Christians in the world. Also, Ellul finds in the Christian faith a standpoint from which all social facts can be deconstructed, thus putting together the principles of a new iconoclasm. The same is not true for Berger.

The study of these two authors' nonconformist stances toward the relation between Christian revelation and society can be continued by analyzing their views about technology and modern consciousness. Such a dialogue between the ideas of Jacques Ellul and Peter L. Berger can provide fruitful results for all those interested in the mutation of human mentality brought into being by the modern world.

References

Berger, Peter L. 1969a. *A rumour of angels: Modern society and the rediscovery of the supernatural*. Garden City: Doubleday.

Berger, Peter L. 1969b. *The sacred canopy: Elements of a sociological theory of religion*. Garden City: Doubleday.

Berger, Peter L. 1976. *The precarious vision: A sociologist looks at social fictions and Christian faith*. Westport: Greenwood Press.

Berger, Peter L. 1980. *The heretical imperative: Contemporary possibilities of religious affirmation*. Garden City: Doubleday.

Berger, Peter L. 1998. Protestantism and the quest for certainty. *Christian Century* 115(23): 782–796.

Berger, Peter L. 1999. The desecularization of the world: A global overview. In *The desecularization of the world: Resurgent religion and world politics*, ed. Peter L. Berger, 1–18. Grand Rapids: Eerdmans.

Berger, Peter L. 2001. Reflections on the sociology of religion today. *Sociology of Religion* Winter, 62(4): 443–454.

Berger, Peter L. 2004. *Questions of faith: A skeptical affirmation of Christianity*. Oxford: Blackwell.

Berger, Peter L., and Thomas Luckmann. 1967. *The social construction of reality: A treatise in the sociology of knowledge*. Garden City: Doubleday.

Ellul, Jacques. 1964. *The technological society*. Trans. John Wilkinson. New York: Knopf.

Ellul, Jacques. 1972. *The politics of god and the politics of man*. Trans. Geoffrey W. Bromiley. Grand Rapids: Eerdemans.

Ellul, Jacques. 1973. *The meaning of the city*. Trans. Dennis Pardee. Grand Rapids: Eerdemans.

Ellul, Jacques. 1975. *The new demons*. Trans. C. Edward Hopkin. Oxford: Mowbrays.

Ellul, Jacques. 1980. *The technological system*. Trans. Joachim Neugroschel. New York: Continuum.

Ellul, Jacques. 1983. *Living faith: Belief and doubt in a perilous world*. Trans. Peter Heinegg. New York: Harper & Row.

Ellul, Jacques. 1989a. *The presence of the kingdom*. Trans. Olive Wyon. Colorado Springs: Helmers & Howard.

Ellul, Jacques. 1989b. *The subversion of Christianity*. Trans. Geoffrey W. Bromiley. Grand Rapids: Eerdemans.

Ellul, Jacques. 1989c. *What I believe*. Trans. Geoffrey W. Bromiley. Grand Rapids: Eerdemans.
Goddard, Andrew. 2002. *Living the word, resisting the world: The life and thought of Jacques Ellul*. Bletchley: Paternoster.
Martin, Bernice. 2001. Berger's anthropological theology. In *Peter Berger and the study of religion*, ed. Linda Woodhead, Paul Heelas, and David Martin, 154–188. London/New York: Routledge.
Milbank, John. 1993. *Theology and social theory: Beyond secular reason*. Oxford: Blackwell.
Rowland, Tracey. 2003. *Culture and the Thomist tradition: After Vatican II*. London: Routledge.

Chapter 17
Postmodernity, the Phenomenal Mistake: Sacred, Myth and Environment

Gregory Wagenfuhr

Jacques Ellul in *The New Demons* proposes that one of the three "phenomenal mistakes" of all Christian history is misunderstanding the modern world as secularized (Ellul 1975: 213). Whilst this may be characteristic overstatement, Ellul is correct in seeing the danger of an entirely opposite appraisal of the social situation. Postmodernity, I will argue, is one such phenomenal mistake.[1]

Two points in geometry, as well as in history, are sufficient to draw a line or trajectory. One can evaluate differences between the two points and progress may be said to have occurred. In actual fact, three points are necessary for the plotting of a structure, a shape, and thus an analysis of one specific point. With three, or more, points each point can be understood relative to a third option. In navigation, of course, the knowledge of three points enables one to plot an exact location, whereas two points are insufficient. What is clear in geometry or navigation is unfortunately often overlooked in historical or sociological narration. This is clear in the case of the idea of postmodernity. A trajectory has been established, but the location is quite vague.

Postmodernity, and similar post-narratives, are negative answers to the generally unstated question, "Where are we?". While the *via negativa* is essential in theology, in sociology it is not helpful. The narratives of where-we-are-not or who-we-no-longer-are are of little help in determining where or who we are. The narrative of postmodernity often begins with some generally understood concept, e.g. the industrial age, philosophical modernism, Enlightenment rationalism, the reign of science. The narrative proceeds to a new situation that no longer looks like the initial point. Two points are given, one positive, one negative. By placing the narrative setting in modernity, the ending of the narrative can only be constructed in modern terms.

[1] I, of course, intend the double meaning – a mistake of great proportions as well as a mistake pertaining to the phenomena of society.

G. Wagenfuhr, Ph.D. (✉)
Trinity College, Stoke Bishop, Bristol BS9 1JP, UK
e-mail: greg.wagenfuhr@gmail.com

H.M. Jerónimo et al. (eds.), *Jacques Ellul and the Technological Society in the 21st Century*, Philosophy of Engineering and Technology 13,
DOI 10.1007/978-94-007-6658-7_17, © Springer Science+Business Media Dordrecht 2013

There is either progress or regress, but there is no positive answer to the current human situation. What is needed is a third point so that structure and triangulation may occur. Enter Jacques Ellul.

In this essay, I use Jean Francois Lyotard's *The Postmodern Condition* as an exemplary narrative of postmodernity and I contrast it with Jacques Ellul's human environmental metanarrative found referenced in a number of his works, from *The Technological Society* to *The Technological Bluff*, but most fully explained in *What I Believe* (*WIB*). I argue that Lyotard's narrative creates a myth of postmodernity that enables a self-justification that Ellul's metanarrative finds both naive and dangerous. Instead of simply referencing Ellul's four chapters in *WIB*, I will more broadly consider his perspective on the sacred and myth in relation to his three environments of nature, society and technique. This enables a more detailed metanarrative to result and one that has the potential for the biting self-critique necessarily lacking in a two-point narrative.

1 Lyotard's Postmodernity

Lyotard identifies many features of postmodernity. Whilst it is not feasible to consider all, or even most, of his account here, one aspect of his thesis is essential to the discussion, namely that of narrative and the problem of legitimation. What has defined postmodernity is the realization that modern science, with its suspicion of metaphysical first-principles and final causes, lacks external and objective grounds for legitimation (Lyotard 1984: 29). Because of this, it was inevitable that science turned to the narrative form for legitimation, in spite of its long struggle against this form. Science has adapted and found legitimation in performativity, pragmatics, and consensus. Even in these cases, however, narrative underlies and ultimately is the means of legitimation. The need of narrative legitimation is essential here. The difference between postmodernity and modernity is perhaps not so much one of method of legitimation, but the scope of narrative legitimation. Thus, Lyotard says toward the end of his account,

> We no longer have recourse to the grand narratives – we can resort neither to the dialectic of Spirit nor even to the emancipation of humanity as a validation for postmodern scientific discourse. But as we have just seen, the little narrative (*petit récit*) remains the quintessential form of imaginative invention, most particularly in science (1984: 60).

Thus, the key point that many have assumed in the narration of postmodernity is the suspicion of metanarrative, and the self-aware legitimation through little narratives, however uncomfortable many are with such legitimation.

Lyotard's account, correct as it is in many ways, is ultimately an account moving from a known state of modernity to a new state that is not modernity but is also surely not premodern. It is a story that describes a people who are actively searching for an answer to the question of legitimacy, or justification. The bald assertion of power without need of justification has proven unacceptable, time and again. As Ellul says, "We know that *power always destroys values and meaning*" (Ellul and Vanderburg

2004: 40, italics in the original). Naked power must be clothed in the wondrous vestments of legitimating narrative. Power no longer wears the singular grand robe and crown of the king but is seemingly democratized, wearing the garments of little narratives.

Thus, postmodernity has given rise to a particular kind of diversity, a diversity where each can express one's own narrative of legitimation and is thereby justified. But this diversity is surface level, ephemeral, and these little narratives are myth as I will attempt to show. Before I do so, however, it is necessary to introduce Ellul's account of the current situation.

2 Ellul's Three Environments

In a world critical of the metanarrative, Ellul offers just such a metanarrative in *WIB*. He devotes four chapters to the description of a grand narrative of all of human history, what he calls "The Human Adventure." Ellul presents an environmental metanarrative investigating three epochs that he calls the "prehistoric," "historic" and "posthistoric." These three eras are each characterized by the dominance of a certain kind of environmental factor: nature, society and technique, respectively.

Ellul's notion of environment is characterized by three points, two of which are contradictory:

> [1] On the one side, we find in our environment all that we need to live. I say to *live* and not just survive. We are set in it and we adapt to it, but we also try to adapt it to us.... [3] Also, in relation to the environment we have occasion to exercise one of the most basic functions of life, that is, symbolism. The environment gives us the chance to create symbols, and here are the riches that spur us to development.... [2] But if we really want to understand the environment in its totality, we have to take into account what I have called the contradictory factor. The environment is what puts us in danger. It is both helpful and hostile. When we die, it is always because of the environment (Ellul 1989b: 99–100).

The third factor, Ellul adds, is that of immediacy. All factors in life are mediated through the environment. Thus, environment is (1) the condition of life, (2) the source of death and (3) that which mediates all aspects of life and thus provides the content to symbols.

Ellul's narrative begins with prehistory wherein humanity's environment was nature. Society and technique existed simultaneously, but they were submitted to and mediated through nature. With the settling of cities, the rise of agriculture and the emergence from the neolithic period, humanity entered the social environment where nature and technique were mediated through society. Finally, and most importantly, Ellul thinks that humanity has moved beyond the social environment to the technical environment through which nature and society are mediated. This entry into the technical environment, then, is Ellul's appraisal of the contemporary human situation found in most of his works. Human techniques are not new phenomena for Ellul, what is radically new and different is its role in mediating and ultimately forming the raw material of human experience.

Rather than simply a change in legitimation and narrative, humanity has entered a new environment that follows its own internal laws, laws that are very different from those that reigned in the social milieu. Lyotard himself identifies some of these necessities imposed by technology, as have most who look into the post-modern world.

The strength of Ellul's grand environmental narrative lies in its ability to identify that phenomena of change on a micro-scale are, in fact, rather superficial. Because humanity has recently emerged from the social environment, narratives that are rooted in that environment are sure to end in a new and uncharted world with grave dangers and new prospects. Lyotard, for example, concludes his account with a few suggestions about how computerization could influence and be better used for the sake of justice (Lyotard 1984: 67). Such an account and hope occur from the perspective of the social environment. The perceived need is for justice in the face of the possible terror of technology. Technology must be humanized. But humanity is no longer primarily a social being in Ellul's account. Therefore, the question must not be "How can we use technology for justice?" Justice is no longer an end.[2] Indeed, there are no more ends (Ellul 1964: 19). The question one must ask from Ellul's account is, "What is the social and natural order necessitated by technique?". A question to which Ellul devotes volumes of answers. Nearly every book Ellul published, from his initial *Theological Foundation of Law* that proposed that "technical law" came with the dissolution of a social group (Ellul 1946: 31–36),[3] to his final books of the late 1980s (in French), at least contains the notion of *la technique*, though most have a section dealing with the necessities of this new environment. Many of his specifically Christian books arise as an attempt to answer the similar question, "What does faith look like in the technical environment?" For example, he asks about the Word of God in an image-centric world (Ellul 1985), prayer for the modern person (Ellul 1970), hope in an age seemingly abandoned by God (Ellul 1973), meaningful action in a Christian world of inaction (Ellul 1989a) or action for its own sake (Ellul 1972). All of these publications investigate Christian faith in a newly technical world.

This is the crucial point and the strength of Ellul's metanarrative, as it takes this environmental perspective into account. The technical environment is a specific perspective that people now operate within. It is a worldview in the literal sense of the term – a perspective from which the world is seen. Indeed, the environment forms the basis on which the world can be explained. The environment provides the symbolic content necessary for a linguistic construction of the world.[4] Narratives that begin in modernity, however, still operate within the social perspective and thus

[2] "Unfortunately, efficiency is a fact and justice a slogan" (Ellul 1964: 282).

[3] Indeed, the seeds of *la technique* are clearly evident in his preliminary chapter.

[4] Thus, the transition between the social and technical environment can be seen in that technology has heretofore used social and natural metaphors: e.g. "friends," "tweet," "cloud," "web." But some of these metaphors have begun to reverse, in that the primary meaning is technical and the social or natural meaning increasingly becomes metaphorical: e.g. "network" as referring to social relationships, the understanding of nature as an "eco*system*," viewing people as "capital," or as a "human resource."

fail to take into account the monumental shift that is now occurring. But if these modernity-to-postmodernity accounts fail to identify the perspective from which humans operate, the third point in triangulation, they are very effective as myth, a point for which we must look at the sacred to understand.

3 Ellul, the Sacred and Myth as Orientation

Ellul's environmental account from *WIB* is remarkably similar to that in *New Demons* (*ND*) where Ellul describes the sacred as orientation to space, time and society. The sacred, Ellul thinks, is a "veritable topography of the world," and is therefore "a bestower of meaning" (1975: 52). Ellul also in *New Demons*, quoting Roger Caillois, calls the sacred "the condition of life *and* gate of death" (Ellul 1975: 57; Caillois 2001). Thus, his three descriptive points of the environment in *WIB*, (1) the condition of life, (2) the source of death, and (3) that which mediates experience, are near equivalents to those described in *New Demons* as the sacred. That the sacred orients both individuals and social groups to the environment is important to recognize for Ellul thereby brings a spiritual element into his grand narrative.

Whilst the sacred is not Ellul's most widely recognized field of study, nor is it his greatest, it has a larger importance than its somewhat brief treatment in *New Demons*, *Subversion of Christianity* and a few select articles would suggest. If environment nearly corresponds to Ellul's idea of necessity, as environment is the greatest necessary condition, the sacred brings in a spiritual and human aspect to this situation. The environment, like Ellul's "necessity" is not fate (1976: 37 ff., 1971: 233 ff.). Humans are solely responsible for the transition from nature to society to environment, but there is a "plus factor" of the Powers, as Ellul says (1986: 175). The Powers, which I take to be reifications of what people sacralize, cannot exist apart from people, but have immense impact upon people. Their existence is the result of a general willing them into existence, but that willing is spontaneous and unconscious. The Powers create necessity, but they in turn are created by humanity. This dialectic is equivalent to the orientation to the environment that Ellul describes in *WIB*:

> When a human group gives itself a form that implies the existence of a power or authority, a process of compensation begins. This is always true in history, and I have no hesitation projecting it back into prehistory. In society no force has ever been at work without giving rise to a counterforce.... Here again we must avoid two extremes. On the one hand, this creation is not an organic, spontaneous, or automatic one, as though society created its own antibodies. On the other hand, it is not the fruit of reflection, of some philosophy or theory. Humanity stands between the two. There is a voluntary creation, but only as dictated by circumstances (1989b: 112).

Thus, humans integrate themselves into the environment, creating the Powers of necessity in terms drawn from the environment. This orientation is ambiguous, it is not conscious or rational, nor is it unwilled and fatalistic. Ellul puts it similarly in *ND*:

> Man's movement toward sacralization has its source in his relations with the universe. In a world which is difficult, hostile, formidable, man (unconsciously, spontaneously, yet willingly,

> to be sure) attributes sacred values to that which threatens him and to that which protects him, or more exactly to that which restores him and puts him in tune with the universe (1975: 50).

The creation of social structure is an important feature of the sacred. The creation of these structures is modelled precisely on orienting people to that which most threatens and that which most restores, that which is sacralized, that which is the greatest necessary condition: the environment. And it is the role of myth to effect this orientation through the power of symbolic and simplified narrative.

Ellul takes a far more functional and structural approach to myth than a phenomenological one. For Ellul, an overarching definition of myth that encompasses phenomena is impossible because myth must be expressed in terms related to the environment. Ellul thinks that,

> One all-embracing definition of myth robs it of just that which makes it a myth. According to this, a myth is the interpretation of a very direct relationship between man and the temporal structure of his life. Outside that relationship his life is dust and absurdity. It doesn't seem to me that any overall definition is possible which would apply equally to our twentieth-century myth and those of three thousand years ago.... If myth is a mirror of man's reflection, if it is an explanation of man's action, if it is a grasp on and a justification of man's situation *hic et nunc*, if, finally, it is an image of the most mysterious depths of man in confrontation with a given reality, then it cannot, by its very nature, be the same now as then. Myth necessarily appears in specific forms, but its characteristics and reasons are constant and common to all (1975: 91–92).

In Ellul's account, if there is a static function of myth then its phenomena must be varied depending on the situation in which the myths are expressed.

Myths, I must add, are necessarily variegated and ad hoc. Myths are, by necessity, little narratives. Little narratives are diverse. They do not incorporate an entire explanation of the human experience, but they deal with specific aspects of that experience. Myths, likewise, explain only specific aspects. It is possible to construct a pantheon, sometimes with highly contradictory little narratives, out of a culture's collection of myths. But the elimination of any single myth would not undermine the pantheon. Thus, myths are not comprehensive. In becoming comprehensive, a myth becomes a metanarrative, and metanarratives are dangerous to the sacral order insofar as metanarratives provide the ability to determine limits and thus pursue a definition. The multiplicity of myths allow for an endless ability to attach different myths, to accept contradictory myths, to believe in the rightness of specific phenomena without reference to a grand story whose limits are known. Metanarrative is self-conscious limitation. Conscious limitation implies conscious definition, which in turn implies that what has been defined is not self-evident. A metanarrative is never self-evidently true, for it takes into account features of other places and times. Myth, however, works precisely because it assumes, and reinforces, "self-evident" truths.[5] Only in this manner can myth provide the function of justification

[5] I use "self-evident" in a relativist sense. That is, truths that are self-evident are socially relative and are indicators of the sacred. They point to *a priori* beliefs that are often unconscious.

without the hearers becoming critical of the structure of the little narratives, for no structure is assumed to exist apart from reality. If I can be so bold as to advance my own functional definition of myth here, it is *legitimation through little narratives.*

This legitimation through little narratives is visible in the mythological library of every society. From stories of the Battle of Britain Spitfire aces to Martin Luther King Jr.'s famous speeches and death, to the late Steve Jobs, the genius who humanized the computer, all have a role in constructing an identity of an "us" and a "them." Myths need not be fabrications or "untrue," nor do they need not incorporate divinity (Ellul 1975: 92), but they always present a subtle, though partial, portrait of a social identity that should be revered, imitated and sanctified. Thus, myths favor extremely specific elements of a historical narrative that serve a contemporary justificatory purpose. But it would be a mistake to assume that myth is only related to social integration and identity purposes. Aetiological myths provide legitimation for environmental facts, e.g. why the zebra has stripes to a people in the natural environment, why the king is to be revered in a social environment, or why progress is necessary and good in the technical environment. Thus, as Ellul says, "as is the case with the sacred, the domain of myth is shifted. It no longer refers to nature (cosmogony) but to the real problems of the culture of our day" (1975: 94).

4 Postmodernity as a Myth

The aspect of justification or legitimation is the static function of myth, whatever literary forms or content it takes. It must be recalled that, for Ellul, there can be few more dangerous acts than the justification of necessity, i.e. orientation to environment. Ellul says this starkly in *Ethics of Freedom*,

> Spiritually the most destructive and deceptive act is that of making a virtue of necessity. Obeying determinations is never a virtue even if it leads to success... or the good (1976: 45–46).

Thus, for Ellul, myth is categorically spiritually destructive. And, if my definition of myth as legitimation through little narratives is acceptable, it is apparent that postmodernity is, in fact, a period of human history drowning in myth and the glorification of myth. It is, in an Ellulian reading, a spiritually dangerous time.

I mentioned above that the apparent diversity of postmodernity is often its most praised aspect. Each individual is said to have the right to choose any number of beliefs, whether they be religious, sexual, or cultural. This is a spiritually dangerous myth for it carefully masks power. The right to choose what one believes is a false situation. Belief in the sacred structures, in the orientation to reality, is not one's choice but is the very perspective from which one explains all reality. Perspective is presuppositional on the linguistic level. The choices, then, that people are socially entitled to make are superficial. They belong quite clearly to the social environment. Sexuality, marriage, traditional religions, cultural peculiarities, all these are irrelevant to the technical environment, so long as they do not impede efficiency and stability. The diversity of beliefs and of narratives is not a feature of a new and free humanity, but, ironically, evidence of a radically integrated humanity. That people can live in relative harmony while holding what would traditionally be socially intolerable differences is not a victory to human understanding, but a tacit admission that culture is now superficial and a nonessential aspect of personal and social identity. Indeed, culture is now a consumer product.

These myths of postmodernity follow the two-pointed narrative I described above. Because we think we know what modernity was, and because we are no longer fooled by metanarratives, we believe we live in a period of great freedom and toleration. Postmodernity is narrated as a period of transition, coming out of a rigidly scientific era, embarking on a new project whose direction and destination is uncertain. This narrative justifies the situation as one of progress, freedom, possibility.

Ellul's environmental metanarrative, however, provides a radical critique of this postmodern narrative. Rather than postmodernity being a time of transition, postmodernity is, in fact, a period of great integration. The technical environment became increasingly dominant in the twentieth century. Ellul describes this well in his trilogy on technique, there is no need here to reiterate his most well-known theory.[6] But Ellul begins *The Technological System* by saying that "Twenty-five years ago, I arrived at the notion of the 'technological society'; but now, that stage has passed" (1980: 1). It has now been nearly 35 years since Ellul wrote that sentence, and almost six decades from (Ellul 1954). The period of transition is over. The technical environment reigns supreme and society is producing the little narratives that justify the situation while masking its realities. As long as the post-narratives continue, as in postmodernity, we continue to integrate ourselves into the technical environment and all that it entails.

Make no mistake, the seeming successes of what might initially be understood as anti-technical movements, e.g. environmentalist movements, are also myths that must be combated on the same grounds of environment. The green narrative is told from the perspective of the technical environment. "Being green" has been equated with recycling, with driving fuel efficient vehicles, with riding bicycles more. All of these activities are technical activities designed to increase efficiency, that is, to slow the consumption of nonrenewable resources, to prevent global climate change, which potentially hinders growth, and to protect the possibility of a technological future. The stability of technical progress is the underlying motive of much of the green movement, the preservation of the *status quo* in light of the dangers of overconsumption, and not a prioritisation of nature, or human society, over technique. Thus, the keyword is "sustainability." The technical world is defined by two factors: "consciousness and judgment" (Ellul 1964: 20). The very fact that our

[6]Many interpreters of Ellul have confined themselves to (1954) and (1964) and have misunderstood the differences between technology itself: society under the effects of technology, and technique as *the* factor of a radically new historical epoch. The latter is Ellul's ultimate point. Any analysis of Ellul's concept of *la technique* that misses this point is partial at best, and is usually misguided, unfair and thus coming to poor conclusions.

society must consciously attempt to preserve the natural world demonstrates the technical perspective. As long as nature is the "great *outdoors*," i.e. a place outside the sphere of daily human activity, as long as people in their work are called a "human resource", as long as the term "community" can legitimately be applied to people who will never have physical contact, any supposedly anti-technical movement will either be subverted from the beginning (as is the case with many hard-core environmentalist movements or social movements like religion), or will be propagandistic (as is the case with governments and large corporations).

Other post-narratives demonstrate similar mythical qualities, especially the complete inability for self-criticism. Post-colonialism is another popular post-narrative that is also a phenomenal mistake. National colonialism is mostly dead, whilst technological, corporate and even paternalistic charitable colonialism lives and thrives. It is ironic that in a world that congratulates itself on its diversity of little narratives there is actually an underlying *a priori* sacred belief in a definition of what-it-means-to-be-human that morally necessitates the "evangelizing" of the "savages" with the "gospel" of modern medicine, American food franchises, and the human right to Internet access. This colonial attitude is a necessary aspect of the sacred, necessary not only to legitimate one's own worldview, but also for to the true good intention of civilizing the savage. Mircea Eliade explains that, at least for ancients, though it seems similarly true for moderns, what is foreign is chaotic and must be consecrated, which is to say, created and turned into cosmos (Eliade 1959: 29–32). Those who do not exist in the technical environment do not yet exist as fully human,[7] and it is a moral indictment against us if we are to call them "human" and do not provide them the necessary means to become human.[8] This is nothing more than masking technical colonialism behind social myths of justice and equality.

5 Conclusion: Consequences

The consequences of the continuation of narratives of postmodernity and the legitimation this brings to little narratives, especially those of the apparent social diversity, are potentially dire. Ellul notes that a certain diversity is promoted by technique:

> Two cultures, of which technique is one, cannot coexist. This does not mean, of course, that uniformity prevails…. Technique does not lead to a general uniformity. In fact, it creates a certain diversity. Its objectives are always the same, and so is its influence on man. But

[7]This is an example of an oft committed logical mistake of substituting environmental or external conditions for something non-environmental or internal. The definition of "being human" cannot in itself contain a specific environment. One may say that a human is more likely to thrive in a certain environment, but the environment is not responsible for thriving. The definition of human cannot in itself contain factors external to humanity. The oft mentioned term "quality of life" is an excellent example that substitutes a subjective qualitative judgment of life that each person must personally make for an objective quantity such that the reproduction of high quantity will necessarily lead to high quality.

[8]On an understanding of "being human" by technical means, see Ellul (1990: 213 ff.).

> though it is axiomatic that the one best way will prevail, this one best way will vary with climate, country, and population. The more technique is refined, the more it varies its means of action (1964: 130).

Thus, technique and the technical environment may well produce a variety of little narratives. These little narratives are superficial choices:

> Man can choose. But his choices will always bear upon secondary elements and never on an overall phenomenon. His judgments will always be ultimately *defined* by the technological criteria.... Man can choose, but in a system of options established by the technological progress (1980: 325).

But, the one best way will prevail, not in the mechanization of people into automatons performing equivalent tasks, but in creating a unified language and environment within which people operate. Thus, to celebrate social diversity is not to celebrate the liberty of human choice as much as it is to mask the growing global uniformity necessitated by the laws of technique. National, religious, racial and gender identities are becoming increasingly superficial insofar as a person of any background, creed or gender can operate a computer with equivalent performance. People of different cultures and languages use the same technological gadgets that utilize the same internal and specialized language to mediate basic life activities. This linguistic uniformity evidences the rise of a new, increasingly uniform environment.

This brief argument has covered many concepts to argue that postmodernity is a dangerous myth, and that Ellul provides a metanarrative that enables critique of this myth from a transhistorical perspective. I am convinced that the most dangerous power humans wield is that of justification. At the same time, humanity's greatest weakness is its need for legitimation. Myth and the sacred are precise methods that enable humans to exercise the power of justification without the godlike responsibility that self-justification entails. Ellul recognized the danger of myth and the sacred in *New Demons*. The phenomenological forms that the sacred and its myths take have changed since the 1970s, let alone since ancient Greece. Myth and the sacred, however, have not died away.

Furthermore, because technique necessarily creates a deep environmental uniformity, there is a growing danger that the sacred and its myths will be impossible to resist. What was peculiar about previous eras of human history was the diversity of environments experienced. During the social environment, there were always groups of people close to nature who were radically different. And this knowledge of great linguistic, religious and mythical diversity had a great demythologizing and desacralizing impact on Western society. The presence of competing metanarratives and worldviews is a healthy situation that requires self-criticism and an expansion of horizons. In seeking to understand different worlds, in learning new languages, one loses confidence in the absolute and self-evident truths that the sacral orders produce. With a growing uniformity, especially the growing influence of English and technical language, voices of criticism through true and critical diversity will be eliminated. When the environment is uniform, the sacral order will be uniform, human symbolic expression will be uniform and thus absolute truth will reign, only it is an absolute truth created, not by arriving at a destination, at God or the Ideal, but by the elimination of dissenting voices. It is truth by consensus, by general corporate legitimation, which is the most cowardly approach to truth of all. Consensual truth is not truth, it is shared opinion mistaken for truth in the absence of any dissent. With no ground on which to critique this shared opinion, with no differing perspective brought about by true and radical difference, the technical environment introduces radical uniformity with no possibility of dissent while maintaining the semblance of diversity and choice.

Indeed, Ellul ends his final book on the phenomenon of *la technique* (1990) with a chapter on "Terrorism in the Velvet Glove of Technology."[9] The term terrorist here refers to an imposition of practice, opinion, belief, and ultimately worldview, upon an individual or social group that has no ability to resist. Ellul, with typical sarcasm, describes the society of the near future (from 1988) in these terrorist terms:

> This totally technicized, computerized society is inevitable. We have to go with the flow, to make it arrive, to preside at its birth, and to integrate the new generation into this world. We no longer have any choice. There are no options, which would be useless, for we know what the outcome will be.... The ineluctable outcome is dictatorship and terrorism. I am not saying that the governments that choose this as the flow of history will reproduce Soviet terrorism. Not at all! But they will certainly engage in an ideological terrorism (1990: 386–387).

This terrorism is not blatant or outwardly violent. It dons the vestments of democracy and public opinion. The majority always goes with the flow of history, since the majority creates the flow of history unconsciously. Ellul describes three threats which form this terrorism: (1) unemployment through technical ignorance; (2) intelligence conceived in technical terms, which leads to the neglect of learning in the humanities; (3) education, in that children are required to adapt to technology, especially computers. It is the pedagogical issues in particular that lead Ellul to call technique "terrorist." He says:

> This is why I talk of terrorism. *All* children must now learn to use computers. They are shaped by them and adapted to them. This adaptation will one day go so far that orthography itself will have to be changed to fit computers. The computer will mediate all things intellectual and the whole intellectual formation of the child (1990: 389, italics in the original).

Ellul is incorrect about orthography insofar as he does not go far enough: orthography is now somewhat obsolete and falling into neglect. But his point about computers mediating all things intellectual is clearly now true. Because this occurs in the formation of a young mind and education is increasingly standardized and submitted to technique, there becomes no alternative. Whereas in the social environment there existed different socioeconomic classes with radically different educations, ideas, values, and dialects within the same society, as well as the radical differences found in other societies, in the technical environment education, though it may have different content, is all performed and mediated through a computerized

[9] Ellul's choice of "terrorism" becomes increasingly unfortunate as the word now seemingly only refers to "radical" idealists, "fundamentalist" Islam, or similar individuals or groups who commit mass acts of violence to communicate a point. I prefer the term "colonialist" or "imperialist," though both of these terms imply a national or statist agency, which is not my point.

environment that imposes a certain linguistic and experiential uniformity whose influence can hardly be underestimated.

It is therefore paramount that Ellul's influence increases in the twenty-first century for any who consider the technical environment distasteful, morally reprehensible, or evil. Whilst his style is often rather specific to his contexts, the concepts behind his social critiques are still highly valid and still dangerous to the *status quo*. Critique of the technical environment in itself is now nearly irrelevant. We live within that environment, we have sacralized it. It feeds us, protects us, and creates the conditions of life. We cannot critique it from within, using its own terms. But Ellul offers far more than just a critique of *la technique*, he offers a perspective and a metanarrative framework from which it is still possible to critique the sacral order justified through myth. As Ellul himself said over a quarter century ago,

> Now, more than ever before, man is enslaving himself to things and to other men through the religious process. It is not technology itself which enslaves us, but the transfer of the sacred into technology. That is what keeps us from exercising the critical faculty, and from making technology serve human development (1975: 206).

This critique is essential. Postmodernity as a narrative of little narratives reveals a period of high myth and sacralization. Its mistake of scope in identifying only two points, and thus only a trajectory, prevents the possibility of criticism at a time when criticism is most vital. Postmodernity is, therefore, a phenomenal and dangerous mistake.

References

Caillois, Roger. 2001. *Man and the sacred.* Trans. Meyer Barash. Urbana: University of Illinois Press.

Eliade, Mircea. 1959. *The sacred and the profane: The nature of religion.* Trans. Willard R. Trask. New York: Harcourt, Brace.

Ellul, Jacques. 1946. *Le fondement théologique du droit.* Neuchâtel: Delachaux & Niestlé.

Ellul, Jacques. 1954. *La technique ou l'enjeux du siècle.* Paris: Armand Colin.

Ellul, Jacques. 1964. *The technological society.* Trans. John Wilkinson. New York: Knopf.

Ellul, Jacques. 1970. *Prayer and modern man.* Trans. C. Edward Hopkin. New York: Seabury Press.

Ellul, Jacques. 1971. *Autopsy of revolution.* Trans. Patricia Wolf. New York: Knopf.

Ellul, Jacques. 1972. *False presence of the kingdom.* Trans. C. Edward Hopkin. New York: Seabury Press.

Ellul, Jacques. 1973. *Hope in time of abandonment.* Trans. C. Edward Hopkin. New York: Seabury Press.

Ellul, Jacques. 1975. *The new demons.* Trans. C. Edward Hopkin. New York: Seabury Press.

Ellul, Jacques. 1976. *The ethics of freedom.* Trans. Geoffrey W. Bromily. Rapids: Eerdmans.

Ellul, Jacques. 1980. *The technological system.* Trans. Joachim Neugroschel. New York: Continuum.

Ellul, Jacques. 1985. *The humiliation of the word.* Trans. Joyce Main Hanks. Grand Rapids: Eerdmans.

Ellul, Jacques. 1986. *The subversion of Christianity.* Geoffrey W. Bromiley. Grand Rapids: Eerdmans.

Ellul, Jacques. 1989a. *The presence of the kingdom.* Trans. Olive Wyon. Colorado Springs: Helmers & Howard.

Ellul, Jacques. 1989b. *What I believe*. Trans. Geoffrey W. Bromiley. Grand Rapids: Eerdmans.

Ellul, Jacques. 1990. *The technological bluff.* Trans. Geoffrey W. Bromiley. Grand Rapids: W.B. Eerdmans.

Ellul, Jacques, and William H. Vanderburg. 2004. *Perspectives on our age: Jacques Ellul speaks on his life and work.* Toronto: House of Anansi.

Lyotard, Jean-François. 1984. *The postmodern condition: A report on knowledge*. Trans. Geoff Bennington and Brian Massumi. Minneapolis: University of Minnesota Press.

Authors

Jennifer Karns Alexander received her Ph.D. from the University of Washington. She is an associate professor in the Program for the History of Science and Technology and the Department of Mechanical Engineering at the University of Minnesota. Her main research interest is technology and modern industrial society, and her current focus is on religion and technology in the post-war era. She is the author of *The Mantra of Efficiency: From Waterwheel to Social Control* (2008), winner of the Edelstein Prize from the Society for the History of Technology.

Andoni Alonso received his Ph.D. from the University of Navarra. He was a Research Fellow at Pennsylvania State University (1996–1998) and University of Nevada Reno (2003). Currently he is associate professor at Universidad Complutense de Madrid. His recent publications include *Diasporas in the New Media Age* (2010, coedited with Pedro Oiarzabal), *Knowledge Communities* (2011, coedited with Javier Echeverría and Pedro Oiarzabal), and *Ciencia expandida, naturaleza común y saber profano* [Expanded science, common nature and profane knowledge] (2011, coauthored with Antonio Lafuente).

Daniel Cérézuelle studied philosophy and social sciences with Jean Brun, Hans Jonas, and Jacques Ellul. As a philosopher he has taught philosophy of technology in France and in the United States and served on the board of the *Société pour la Philosophie de la Technique*. As a sociologist he is investigating the social importance of non-monetary economy in a modern society and serves as Scientific Director of the *Programme Autoproduction et Développement Social* (*PADES*). He has recently published *La technique et la chair* (2011).

José Luís Garcia is a research fellow at the Institute of Social Sciences, University of Lisbon. He received his Ph.D. in social sciences from the same University after pursuing doctoral studies at the London School of Economics and Political Science, University of London. His main research interests are the social theory of technology, sociology of technology, biotechnology, commodification and ethics, communication and information technologies. His publications include *Dilemas da Civilização*

H.M. Jerónimo et al. (eds.), *Jacques Ellul and the Technological Society in the 21st Century*, Philosophy of Engineering and Technology 13,
DOI 10.1007/978-94-007-6658-7, © Springer Science+Business Media Dordrecht 2013

Tecnológica [Dilemmas of the technological civilization] (2003), *Razão, Tempo e Tecnologia* [Reason, time and technology] (2006). He is currently writing a book, *Technodicy*.

Randolph Haluza-DeLay received his Ph.D. in education from the University of Western Ontario. He is an associate professor of sociology at The King's University College in Edmonton. He is lead co-editor of the anthology *Speaking for Ourselves: Environmental Justice in Canada* (2009). Recent research is primarily in environmental sociology, with a focus on environmental justice, social movements, cultures of sustainability, and the sociology of climate change. He has more than two dozen publications in peer-reviewed journals and scholarly book chapters. A book on climate change and global religions is in the works.

Yuk Hui received his Ph.D. from Goldsmiths, University of London, with a thesis on philosophy titled *On the Existence of Digital Objects*. He is currently a postdoctoral researcher at the Institut de Recherche et d'Innovation du Centre Pompidou. His main research interests are in phenomenology, philosophy of technology and computation.

Andrei Ivan received his master's degree from University of Bucharest, Romania. His interests are in the sociology of religion, theology and technology studies.

Helena Mateus Jerónimo received her Ph.D. from the University of Cambridge. She is an assistant professor of the School of Economics and Management (ISEG), Technical University of Lisbon, and researcher at the Research Centre in Economic and Organizational Sociology (SOCIUS). Her main research interests are in science and technology studies; environment, risk and uncertainty; and democracy and citizenship. She has recently published *Queimar a Incerteza: Poder e Ambiente no Conflito da Co-Incineração de Resíduos Industriais Perigosos* [*Burying Uncertainty: Power and environment in the conflict of co-incineration of hazardous industrial waste*] (2010). She is co-editor of the Portuguese journal *Análise Social*.

Nathan Kowalsky received his Ph.D. in philosophy from the Katholieke Universiteit Leuven. He is an assistant professor of philosophy at St. Joseph's College, University of Alberta, as well as assistant professor of science, technology and society and adjunct professor of religious studies through the Office of Interdisciplinary Studies. His main research interests are in environmental philosophy, philosophy of religion, and philosophy of culture – whilst also dabbling in philosophy of technology. His recent publications include an edited volume *Hunting – Philosophy for Everyone: In Search of the Wild Life* (2010) and a paper on social construction, Continental philosophy of religion, and the epistemic status of science (*Zygon*, March 2012).

Isabelle Lamaud is a doctoral student in political sciences at Université de Bordeaux IV. Her main research interests are in ecological political thought, critiques of technical progress and industrialism. Her dissertation is on the work of André Gorz.

Virginia W. Landgraf has been an indexer-analyst at the American Theological Library Association in Chicago since 2004. She wrote her Ph.D. dissertation, *Abstract Power and the God of Love: A Critical Assessment of the Place of Institutions in*

Jacques Ellul's Anthropology of Dialectical Relationships at Princeton Theological Seminary (2003). A board member of the International Jacques Ellul Society, she has published in the *Journal of the Society of Christian Ethics*, the *Ellul Forum*, and *Réforme*. In addition to Jacques Ellul, her research interests include economic ethics and our relationship to time as a spiritual issue.

Carl Mitcham is professor of Liberal Arts and International Studies, Director of the Hennebach Program in the Humanities, and Co-Director of the Ethics Across Campus program at the Colorado School of Mines. Recent publications include the *Encyclopedia of Science, Technology, and Ethics* (2005, 4 volumes), *Oxford Handbook of Interdisciplinarity* (2010, coedited with Robert Frodeman and Julie Thompson Klein), and *Ethics and Science* (2012, coauthored with Adam Briggle).

George Ritzer is Distinguished University Professor at the University of Maryland. Among his books in metatheory are *Sociology: A Multiple Paradigm Science* (1975/1980) and *Metatheorizing in Sociology* (1991). In the application of social theory to the social world, his books include *The McDonaldization of Society* (7th ed., 2013), *Enchanting a Disenchanted World* (3rd ed. 2010), and *The Globalization of Nothing* (2nd ed., 2007). He edited the *Blackwell Companion to Major Social Theorists* (2000), *The Blackwell Companion to Globalization* (2008) and co-edited the *Handbook of Social Theory* (2001). He also edited the 11-volume *Encyclopedia of Sociology* (2007) and the two-volume *Encyclopedia of Social Theory* (2005) and the *Encyclopedia of Globalization* (2012).

Frédéric Rognon received his Ph.D. from the University of Paris X. He is a professor of philosophy in the Faculty of Protestant Theology of the University of Strasbourg. His main research interests are in the thought of Kierkegaard, Bonhoeffer and Levinas, dialectics between faith and reason, and critics of technological society. He has published *Jacques Ellul. Une pensée en dialogue* (2007 and 2012), *Dietrich Bonhoeffer* (2011), and *Générations Ellul. Soixante portraits de la pensée de Jacques Ellul* (2012).

Wha-Chul Son received his Ph.D. from the Catholic University of Leuven. He is teaching philosophy as assistant professor at Handong Global University in Pohang, Korea. His research area is philosophy of technology focused on the democratization of technology, theories of classical philosophers of technology, the technological divide, engineering ethics, and appropriate technology.

Patrick Troude-Chastenet is professor of political science at the University of Bordeaux and a researcher at the Centre Montesquieu de Recherches Politiques (CMRP GRECCAP). His main research interests are in democratic theory, French political history and political ideas, especially those concerning personalism and ecological movements. He has published ten books on Jacques Ellul including *Jacques Ellul, penseur sans frontière* (2005), *Jacques Ellul on Politics, Technology, and Christianity* (2005), and *La Politique* (2008).

Gregory Wagenfuhr is finishing his Ph.D. in theology at the University of Bristol. He is looking at the question of the human need of moral justification with a dialectic of sociology and theology, concluding that Christian faith is life without the need of moral justification. His thesis utilizes Ellul as a primary interlocutor. His interests are in placing theology, philosophy, and sociology in dialogue. He is also a fine art landscape photographer who seeks to demonstrate the ambiguity of symbol and visual meaning through his work (www.authorityphotography.com).

Langdon Winner received his Ph.D. from the University of California, Berkeley, in political science. He is the Thomas Phelan Chair of Humanities and Social Sciences at Rensselaer Polytechnic Institute. Author of *Autonomous Technology* (1977), *The Whale and the Reactor* (1984), and other works, his scholarly interests include democratic political theory, philosophy of technology, and contemporary political movements. His blog, *Technopolis*, offers occasional commentaries on technology and everyday life. At present he is writing a book, *Power Fantasy: The Rise and Fall of Techno-Triumphalism.*

Index

H.M. Jerónimo et al. (eds.), *Jacques Ellul and the Technological Society in the 21st Century*, Philosophy of Engineering and Technology 13,
DOI 10.1007/978-94-007-6658-7, © Springer Science+Business Media Dordrecht 2013

MIX
Papier aus verantwortungsvollen Quellen
Paper from responsible sources
FSC® C105338

If you have any concerns about our products,
you can contact us on
ProductSafety@springernature.com

In case Publisher is established outside the EU,
the EU authorized representative is:
Springer Nature Customer Service Center GmbH
Europaplatz 3, 69115 Heidelberg, Germany

Printed by Libri Plureos GmbH
in Hamburg, Germany